TEACHER GUIDE
3rd–8th Grade

God's Design: Life

Author: Richard & Debbie Lawrence

Design: Diane King

Editor: Gary Vaterlaus

Master Books Creative Team:

Editor: Craig Froman

Cover Design: Diana Bogardus

Copy Editor: Judy Lewis

Curriculum Review:
Kristen Pratt
Laura Welch
Diana Bogardus

First printing: May 2018
Ninth printing: September 2023

Copyright © 2018 by Richard and Debbie Lawrence. All rights reserved. No part of this book may be reproduced, copied, broadcast, stored, or shared in any form whatsoever without written permission from the publisher, except in the case of brief quotations in articles and reviews. For information write:

Master Books, P.O. Box 726, Green Forest, AR 72638

Master Books® is a division of the New Leaf Publishing Group, LLC.

ISBN: 978-1-68344-128-1
ISBN: 978-1-61458-652-4 (digital)

Unless otherwise noted, Scripture quotations are from the New King James Version of the Bible.

Printed in the United States of America

Please visit our website for other great titles: www.masterbooks.com

Permission is granted for copies of reproducible pages from this text to be made for use within your own homeschooling family activities. Material may not be posted online, distributed digitally, or made available as a download. Permission for any other use of the material must be requested prior to use by email to the publisher at info@nlpg.com.

Author Bio: The God's Design Science series is based on a biblical worldview and reveals how science supports the biblical account of creation. **Richard and Debbie Lawrence,** authors of the series, have a long history of enjoying science. They have both worked as electrical engineers and now Debbie teaches chemistry and physics at a homeschool co-op. While homeschooling their children for 16 years, there was almost always a science experiment going on in the kitchen. Today that tradition is being continued with the next generation as the grandkids enjoy Grandma Science Day once a week.

> "Your reputation as a publisher is stellar. It is a blessing knowing anything I purchase from you is going to be worth every penny!
> —Cheri ★★★★★
>
> "Last year we found Master Books and it has made a HUGE difference.
> —Melanie ★★★★★
>
> "We love Master Books and the way it's set up for easy planning!
> —Melissa ★★★★★
>
> "You have done a great job. MASTER BOOKS ROCKS!
> —Stephanie ★★★★★
>
> "Physically high-quality, Biblically faithful, and well-written.
> —Danika ★★★★★
>
> "Best books ever. Their illustrations are captivating and content amazing!
> —Kathy ★★★★★

Affordable
Flexible
Faith Building

Table of Contents

Welcome to God's Design 5

Integrating the Seven C's 11

Daily Suggested Schedule 15

The World of Plants Worksheets 23

The Human Body Worksheets 113

The World of Animals Worksheets 209

The World of Plants Quizzes and Final Exam 307

The Human Body Quizzes and Final Exam 325

The World of Animals Quizzes and Final Exam 343

Answer Keys

    The World of Plants Worksheets 363

    The Human Body Worksheets 375

    The World of Animals Worksheets 387

    The World of Plants Quizzes 397

    The Human Body Quizzes 401

    The World of Animals Quizzes 405

    The World of Plants Final Exam 409

    The Human Body Final Exam 411

    The World of Animals Final Exam 413

Appendices/Master Supply List 417

Glossaries

    The World of Plants 429

    The Human Body 433

    The World of Animals 437

Note: Quizzes and Tests

This course contains both quizzes and tests to help assess the student's mastery and understanding of key concepts. These assessments also have a second section that includes questions for older students who do the associated lesson challenges. The quizzes and tests have a suggested point system, but as always, you can alter, adjust, or modify these assessments to fit the needs and abilities of your student. The assessments can be given orally as well.

Welcome to *God's Design for Life Teacher Guide*! This exciting course has been enhanced with the following features:

- Combined daily schedule for a one-year course (includes material for The World of Plants, The Human Body, and The World of Animals)

- Each individual lesson sheet on its own page

- Contains a master supply list for each section (plants, human body, and animals) as well a supply list for each lesson — all with convenient checkboxes

- Additional instructions and formatting on the lesson sheets make it easier for the student to follow and the parent to grade

- The icons used in the student book have been added on the lesson sheets as well

- A point system has been assigned to quizzes and tests

- Includes separate answer keys for worksheets, quizzes, and tests for each section (plants, human body, animals)

- Answer keys contain both questions and answers for convenience

- Includes copies of the glossary from the student book

Welcome to GOD'S DESIGN®

LIFE

God's Design for Life is a book that has been designed for use in teaching life science to elementary and middle school students. It is divided into three sections: *The World of Plants*, *The World of Animals*, and *The Human Body*. The course has 105 lessons including a final project that ties all of the lessons together.

In addition to the lessons, special features per section include biographical information on interesting people as well as fun facts to make the subject more fun.

Although this is a complete curriculum, the information included here is just a beginning, so please feel free to add to each lesson as you see fit. A resource guide is included in the appendices to help you find additional information and resources. A list of supplies needed is included at the beginning of each lesson, while a master list of all supplies needed for the entire course can be found in the appendices.

Answer keys for all review questions, worksheets, quizzes, and the final exam are included. If you wish to get through *God's Design: Life* in one year, plan on covering approximately three to four lessons per week. The time required for each lesson varies depending on how much additional information you include, but plan on about 40 to 45 minutes. A helpful daily schedule starts on page 15. Quizzes may be given at the conclusion of each unit and a final exam may be given at the completion of each section.

If you wish to cover the material in more depth, you may add additional information and take a longer period of time to cover all the material, or you could choose to do only one or two of the sections as a unit study.

Why Teach Life Science?

Maybe you hate science or you just hate teaching it. Maybe you love science but don't quite know how to teach it to your children. Maybe science just doesn't seem as important as some of those other subjects you need to teach. Maybe you need a little motivation. If any of these descriptions fits you, then please consider the following:

It is not uncommon to question the need to teach your kids hands-on science in elementary school. We could argue that the knowledge gained in science will be needed later in life in order for your children to be more productive and well-rounded adults. We could argue that teaching your children science also teaches them logical and inductive thinking and reasoning skills, which are tools they will need to be more successful. We could argue that science is a necessity in this technological world in which we live. While all of these arguments are true, not one of them is the real reason that we should teach our children science. The most important reason to teach science in elementary school is to give your children an understanding that God is our Creator, and the Bible can be trusted. Teaching science from a creation perspective is one of the best ways to reinforce your children's faith in God and to help them counter the evolutionary propaganda they face every day.

God is the Master Creator of everything. His handiwork is all around us. Our Great Creator put in place all of the laws of physics, biology, and chemistry. These laws were put here for us to see His wisdom and power. In science, we see the hand of God at work more than in any other subject. Romans 1:20 says, "For since the creation of the world His invisible attributes are clearly seen, being understood by the things that are made, even His eternal power and Godhead, so that they [men] are without excuse." We need to help our children see God as Creator of the world around them so they will be able to recognize God and follow Him.

The study of life science helps us understand the balance of nature so that we can be good stewards of our bodies, the plants, and the animals around us. It helps us appreciate the intricacies of life and the wonders of God's creation. Understanding the world of living things from a biblical point of view will prepare our children to deal with an ecology-obsessed world. It is critical to teach our children the truth of the Bible, how to evaluate the evidence, how to distinguish fact from theory and to realize that the evidence, rightly interpreted, supports biblical creation, not evolution.

It's fun to teach life science! It's interesting, too. Children have a natural curiosity about living things, so you won't have to coax them to explore the world of living creatures. You just have to direct their curiosity and reveal to them how interesting life science can be.

Finally, teaching life science is easy. It's all around us. Everywhere we go, we are surrounded by living things. You won't have to try to find strange materials for experiments or do dangerous things to learn about life.

How Do I Teach Science?

In order to teach any subject you need to understand how people learn. People learn in different ways. Most people, and children in particular, have a dominant or preferred learning style in which they absorb and retain information more easily.

If a student's dominant style is:
Auditory
He needs not only to hear the information but he needs to hear himself say it. This child needs oral presentation as well as oral drill and repetition.
Visual
She needs things she can see. This child responds well to flashcards, pictures, charts, models, etc.
Kinesthetic
He needs active participation. This child remembers best through games, hands-on activities, experiments, and field trips.

Also, some people are more relational while others are more analytical. The relational student needs to know why this subject is important, and how it will affect him personally. The analytical student, however, wants just the facts.

If you are trying to teach more than one student, you will probably have to deal with more than one learning style. Therefore, you need to present your lessons in several different ways so that each student can grasp and retain the information.

Grades 3–8

The first part of each lesson should be completed by all upper elementary and junior high students. This is the main part of the lesson containing a reading section, a hands-on activity that reinforces the ideas in the reading section (blue box) of the student book, and a review section that provides review questions and application questions.

Grades 6–8

In addition, for middle school/junior high age students, a "Challenge" section that contains more challenging material, as well as additional activities and projects for older students (green box), is found in the student textbook.

We have included periodic biographies to help your students appreciate the great men and women who have gone before us in the field of science.

We suggest a threefold approach to each lesson:

Introduce the topic

We give a brief description of the facts. Frequently you will want to add more information than the essentials given in this book. In addition to reading this section aloud (or having older children read it on their own), you may wish to do one or more of the following:

- Read a related book with your students.
- Write things down to help your visual learners.
- Give some history of the subject. We provide some historical sketches to help you, but you may want to add more.
- Ask questions to get your students thinking about the subject.

Make observations and do experiments

- Hands-on projects are suggested for each lesson. This part of each lesson may require help from the teacher.
- Have your students perform the activity by themselves whenever possible.

Review

- The "What did we learn?" section has review questions.
- The "Taking it further" section encourages students to
  - Draw conclusions
  - Make applications of what was learned
  - Add extended information to what was covered in the lesson
- The "FUN FACT" section adds fun or interesting information.

By teaching all three parts of the lesson, you will be presenting the material in a way that children with any learning style can both relate to and remember.

Also, this approach relates directly to the scientific method and will help your students think more scientifically. The *scientific method* is just a way to examine a subject logically and learn from it. Briefly, the steps of the scientific method are:

1. Learn about a topic.
2. Ask a question.
3. Make a hypothesis (a good guess).
4. Design an experiment to test your hypothesis.
5. Observe the experiment and collect data.
6. Draw conclusions. (Does the data support your hypothesis?)

Note: It's okay to have a "wrong hypothesis." That's how we learn. Be sure to help your students understand why they sometimes get a different result than expected.

Our lessons will help your students begin to approach problems in a logical, scientific way.

Icon Key

 Do the activity in the light blue box of your student book (worksheets will be provided by your teacher).

 Test your knowledge by answering the **What did we learn?** questions.

 Assess your understanding by answering the **Taking it further** questions.

 Do the challenge section in the light green box in your student book. This part of the lesson will challenge you to do more advanced activities and learn additional interesting information.

How Do I Teach Creation vs. Evolution?

We are constantly bombarded by evolutionary ideas about living things in books, movies, museums, and even commercials. These raise many questions: Did dinosaurs really live millions of years ago? Did man evolve from apes? Which came first, Adam and Eve or the cavemen? Where did living things come from in the first place? The Bible answers these questions and this book accepts the historical accuracy of the Bible as written. We believe this is the only way we can teach our children to trust that everything God says is true.

There are five common views of the origins of life and the age of the earth:

Historical biblical account	Progressive creation	Gap theory	Theistic evolution	Naturalistic evolution
Each day of creation in Genesis is a normal day of about 24 hours in length, in which God created everything that exists. The earth is only thousands of years old, as determined by the genealogies in the Bible.	The idea that God created various creatures to replace other creatures that died out over millions of years. Each of the days in Genesis represents a long period of time (day-age view) and the earth is billions of years old.	The idea that there was a long, long time between what happened in Genesis 1:1 and what happened in Genesis 1:2. During this time, the "fossil record" was supposed to have formed, and millions of years of earth history supposedly passed.	The idea that God used the process of evolution over millions of years (involving struggle and death) to bring about what we see today.	The view that there is no God and evolution of all life forms happened by purely naturalistic processes over billions of years.

Any theory that tries to combine the evolutionary time frame with creation presupposes that death entered the world before Adam sinned, which contradicts what God has said in His Word. The view that the earth (and its "fossil record") is hundreds of millions of years old damages the gospel message. God's completed creation was "very good" at the end of the sixth day (Genesis 1:31). Death entered this perfect paradise *after* Adam disobeyed God's command. It was the punishment for Adam's sin (Genesis 2:16–17; 3:19; Romans 5:12–19). Thorns appeared when God cursed the ground because of Adam's sin (Genesis 3:18).

The first animal death occurred when God killed at least one animal, shedding its blood, to make clothes for Adam and Eve (Genesis 3:21). If the earth's "fossil record" (filled with death, disease, and thorns) formed over millions of years before Adam appeared (and before he sinned), then death no longer would be the penalty for sin. Death, the "last enemy" (1 Corinthians 15:26), diseases (such as cancer), and thorns would instead be part of the original creation that God labeled "very good." No, it is clear that the "fossil record" formed some time *after* Adam sinned—not many millions of years before. Most fossils were formed as a result of the worldwide Genesis Flood.

When viewed from a biblical perspective, the scientific evidence clearly supports a recent creation by God, and not naturalistic evolution and millions of years. The volume of evidence supporting the biblical creation account is substantial and cannot be adequately covered in this book. If you would like more information on this topic, please see the resource guide in the appendices. To help get you started, just a few examples of evidence supporting biblical creation are given on the following pages.

Evolutionary Myth: Humans have been around for more than one million years.

The Truth: If people have been on earth for a million years, there would be trillions of people on the earth today, even if we allowed for worst-case plagues, natural disasters, etc. The number of people on earth today is about 6.5 billion. If the population had grown at only a 0.01% rate (today's rate is over 1%) over 1 million years, there could be 10^{43} people today (that's a number with 43 zeros after it)! Repopulating the earth after the Flood would only require a population growth rate of 0.5%, half of what it is today.

John D. Morris, *The Young Earth* (Master Books, 1994), pp. 70–71. See also "Billions of People in Thousands of Years?" at www.answersingenesis.org/go/billions-of-people.

Evolutionary Myth: Man evolved from an ape-like creature.

The Truth: All so-called "missing links" showing human evolution from apes have been shown to be either apes, humans, or deliberate hoaxes. These links remain missing.

Duane T. Gish, *The Amazing Story of Creation from Science and the Bible* (El Cajon: Institute for Creation Research, 1990), pp. 78–83.

Evolutionary Myth: All animals evolved from lower life forms.

The Truth: While Darwin predicted that the fossil record would show numerous transitional fossils, even more than 145 years later, all we have are a handful of disputable examples. For example, there are no fossils showing something that is part way between a dinosaur and a bird. Fossils show that a snail has always been a snail; a squid has always been a squid. God created each animal to reproduce after its kind (Genesis 1:20–25).

Ibid., pp. 36, 53–60.

Evolutionary Myth: Dinosaurs evolved into birds.

The Truth: Flying birds have streamlined bodies, with the weight centralized for balance in flight; hollow bones for lightness, which are also part of their breathing system; powerful muscles for flight; and very sharp vision. And birds have two of the most brilliantly-designed structures in nature—their feathers and special lungs. It is impossible to believe that a reptile could make that many changes over time and still survive.

Gregory Parker et al., *Biology: God's Living Creation* (Pensacola: A Beka Book, 1997), pp. 474–475.

Evolutionary Myth: Thousands of changes over millions of years resulted in the creatures we see today.

The Truth: What is now known about human and animal anatomy shows the body structures, from the cells to systems, to be infinitely more complex than was believed when Darwin published his work in 1859. Many biologists and especially microbiologists are now saying that there is no way these complex structures could have developed by natural processes.

Ibid., pp. 384–385.

Since the evidence does not support their theories, evolutionists are constantly coming up with new ways to try to support what they believe. One of their ideas is called punctuated equilibrium. This theory of evolution says that rapid evolution occurred in small isolated populations, and left no evidence in the fossil record. There is no evidence for this, nor any known mechanism to cause these rapid changes. Rather, it is merely wishful thinking. We need to teach our children the difference between science and wishful thinking.

Despite the claims of many scientists, if you examine the evidence objectively, it is obvious that evolution and millions of years have not been proven. You can be confident that if you teach that what the Bible says is true, you won't go wrong. Instill in your student a confidence in the truth of the Bible in all areas. If scientific thought seems to contradict the Bible, realize that scientists often make mistakes, but God does not lie. At one time scientists believed that the earth was the center of the universe, that living things could spring from nonliving things, and that blood-letting was good for the body. All of these were believed to be scientific facts but have since been disproved, but the Word of God remains true. If we use modern "science" to interpret the Bible, what will happen to our faith in God's Word when scientists change their theories yet again?

Integrating the Seven C's

The Seven C's is a framework in which all of history, and the future to come, can be placed. As we go through our daily routines we may not understand how the details of life connect with the truth that we find in the Bible. This is also the case for students. When discussing the importance of the Bible you may find yourself telling students that the Bible is relevant in everyday activities. But how do we help the younger generation see that? The Seven C's are intended to help.

The Seven C's can be used to develop a biblical worldview in students, young or old. Much more than entertaining stories and religious teachings, the Bible has real connections to our everyday life. It may be hard, at first, to see how many connections there are, but with practice, the daily relevance of God's Word will come alive. Let's look at the Seven C's of History and how each can be connected to what the students are learning.

Creation

God perfectly created the heavens, the earth, and all that is in them in six normal-length days around 6,000 years ago.

This teaching is foundational to a biblical worldview and can be put into the context of any subject. In science, the amazing design that we see in nature—whether in the veins of a leaf or the complexity of your hand—is all the handiwork of God. Virtually all of the lessons in *God's Design for Science* can be related to God's creation of the heavens and earth.

Other contexts include:

Natural laws—any discussion of a law of nature naturally leads to God's creative power.

DNA and information—the information in every living thing was created by God's supreme intelligence.

Mathematics—the laws of mathematics reflect the order of the Creator.

Biological diversity—the distinct kinds of animals that we see were created during the Creation Week, not as products of evolution.

Art—the creativity of man is demonstrated through various art forms.

History—all time scales can be compared to the biblical time scale extending back about 6,000 years.

Ecology—God has called mankind to act as stewards over His creation.

Corruption

After God completed His perfect creation, Adam disobeyed God by eating the forbidden fruit. As a result, sin and death entered the world, and the world has been in decay since that time. This point is evident throughout the world that we live in. The struggle for survival in animals, the death of loved ones, and the violence all around us are all examples of the corrupting influence of sin.

Other contexts include:

Genetics—the mutations that lead to diseases, cancer, and variation within populations are the result of corruption.

Biological relationships—predators and parasites result from corruption.

History—wars and struggles between mankind, exemplified in the account of Cain and Abel, are a result of sin.

Catastrophe

God was grieved by the wickedness of mankind and judged this wickedness with a global Flood. The Flood covered the entire surface of the earth and killed all air-breathing creatures that were not aboard the Ark. The eight people and the animals aboard the Ark replenished the earth after God delivered them from the catastrophe.

The catastrophe described in the Bible would naturally leave behind much evidence. The studies of geology and of the biological diversity of animals on the planet are two of the most obvious applications of this event. Much of scientific understanding is based on how a scientist views the events of the Genesis Flood.

Other contexts include:

Biological diversity—all of the birds, mammals, and other air-breathing animals have populated the earth from the original kinds which left the Ark.

Geology—the layers of sedimentary rock seen in roadcuts, canyons, and other geologic features are testaments to the global Flood.

Geography—features like mountains, valleys, and plains were formed as the floodwaters receded.

Physics—rainbows are a perennial sign of God's faithfulness and His pledge to never flood the entire earth again.

Fossils—Most fossils are a result of the Flood rapidly burying plants and animals.

Plate tectonics—the rapid movement of the earth's plates likely accompanied the Flood.

Global warming/Ice Age—both of these items are likely a result of the activity of the Flood. The warming we are experiencing today has been present since the peak of the Ice Age (with variations over time).

Confusion

God commanded Noah and his descendants to spread across the earth. The refusal to obey this command and the building of the tower at Babel caused God to judge this sin. The common language of the people was confused and they spread across the globe as groups with a common language. All people are truly of "one blood" as descendants of Noah and, originally, Adam.

The confusion of the languages led people to scatter across the globe. As people settled in new areas, the traits they carried with them became concentrated in those populations. Traits like dark skin were beneficial in the tropics while other traits benefited populations in northern climates, and distinct people groups, not races, developed.

Other contexts include:

Genetics—the study of human DNA has shown that there is little difference in the genetic makeup of the so-called "races."

Languages—there are about seventy language groups from which all modern languages have developed.

Archaeology—the presence of common building structures, like pyramids, around the world confirms the biblical account.

Literature—recorded and oral records tell of similar events relating to the Flood and the dispersion at Babel.

Christ

God did not leave mankind without a way to be redeemed from its sinful state. The Law was given to Moses to show how far away man is from God's standard of perfection. Rather than the sacrifices, which only covered sins, people needed a Savior to take away their sin. This was accomplished when Jesus Christ came to earth to live a perfect life and, by that obedience, was able to be the sacrifice to satisfy God's wrath for all who believe.

The deity of Christ and the amazing plan that was set forth before the foundation of the earth is the core of Christian doctrine. The earthly life of Jesus was the fulfillment of many prophecies and confirms the truthfulness of the Bible. His miracles and presence in human form demonstrate that God is both intimately concerned with His creation and able to control it in an absolute way.

Other contexts include:

Psychology—popular secular psychology teaches of the inherent goodness of man, but Christ has lived the only perfect life. Mankind needs a Savior to redeem it from its unrighteousness.

Biology—Christ's virgin birth demonstrates God's sovereignty over nature.

Physics—turning the water into wine and the feeding of the five thousand demonstrate Christ's deity and His sovereignty over nature.

History—time is marked (in the western world) based on the birth of Christ despite current efforts to change the meaning.

Art—much art is based on the life of Christ and many of the masters are known for these depictions, whether on canvas or in music.

Cross

Because God is perfectly just and holy, He must punish sin. The sinless life of Jesus Christ was offered as a substitutionary sacrifice for all of those who will repent and put their faith in the Savior. After His death on the Cross, He defeated death by rising on the third day and is now seated at the right hand of God.

The events surrounding the crucifixion and resurrection have a most significant place in the life of Christians. Though there is no way to scientifically prove the resurrection, there is likewise no way to prove the stories of evolutionary history. These are matters of faith founded in the truth of God's Word and His character. The eyewitness testimony of over 500 people and the written Word of God provide the basis for our belief.

Other contexts include:

Biology—the biological details of the crucifixion can be studied alongside the anatomy of the human body.

History—the use of crucifixion as a method of punishment was short-lived in historical terms and not known at the time it was prophesied.

Art—the crucifixion and resurrection have inspired many wonderful works of art.

Consummation

God, in His great mercy, has promised that He will restore the earth to its original state—a world without death, suffering, war, and disease. The corruption introduced by Adam's sin will be removed. Those who have repented and put their trust in the completed work of Christ on the Cross will experience life in this new heaven and earth. We will be able to enjoy and worship God forever in a perfect place.

This future event is a little more difficult to connect with academic subjects. However, the hope of a life in God's presence and in the absence of sin can be inserted in discussions of human conflict, disease, suffering, and sin in general.

Other contexts include:

History—in discussions of war or human conflict the coming age offers hope.

Biology—the violent struggle for life seen in the predator-prey relationships will no longer taint the earth.

Medicine—while we struggle to find cures for diseases and alleviate the suffering of those enduring the effects of the Curse, we ultimately place our hope in the healing that will come in the eternal state.

The preceding examples are given to provide ideas for integrating the Seven C's of History into a broad range of curriculum activities. Give your students, and yourself, a better understanding of the Seven C's framework by using AiG's *Answers for Kids* curriculum. The first seven lessons of this curriculum cover the Seven C's and will establish a solid understanding of the true history, and future, of the universe. Full lesson plans, activities, and student resources are provided in the curriculum set.

AiG offers bookmarks displaying the Seven C's and a wall chart. These can be used as visual cues for the students to help them recall the information and integrate new learning into its proper place in a biblical worldview.

Even if you use other curricula, you can still incorporate the Seven C's teaching into those. Using this approach will help students make firm connections between biblical events and every aspect of the world around them, and they will begin to develop a truly biblical worldview and not just add pieces of the Bible to what they learn in "the real world."

First Semester Suggested Daily Schedule

Date	Day	Assignment	Due Date	✓	Grade
		First Semester-First Quarter			
Week 1	Day 1	The World of Plants Unit 1: Introduction to Life Science Read Lesson 1: Is It Alive? • Pages 14-17 • *God's Design: Life* • (GDL) Complete Worksheet • Pages 25-26 • *Teacher Guide* • (TG)			
	Day 2	Read Lesson 2: What Is a Kingdom? • Pages 18-20 • (GDL) Complete Worksheet • Pages 27-29 • (TG)			
	Day 3	Read Lesson 3: Classification System • Pages 21-23 • (GDL) Complete Worksheet • Pages 31-32 • (TG)			
	Day 4	Read Special Feature: Carl Linnaeus • Pages 24-25 • (GDL)			
	Day 5				
Week 2	Day 6	Read Lesson 4: Plant & Animal Cells • Pages 26-29 • (GDL) Complete Worksheet • Pages 33-34 • (TG)			
	Day 7	Read Special Feature: Cells • Page 30 • (GDL)			
	Day 8	Complete **Introduction to Life Science Quiz 1** (Lessons 1-4) Pages 309-310 • (TG)			
	Day 9	Plants Unit 2: Flowering Plants & Seeds Read Lesson 5: Flowering Plants • Pages 32-34 • (GDL) Complete Worksheet • Pages 35-36 • (TG)			
	Day 10				
Week 3	Day 11	Read Lesson 6: Grasses • Pages 35-37 • (GDL) Complete Worksheet • Pages 37-39 • (TG)			
	Day 12	Read Lesson 7: Trees • Pages 38-40 • (GDL) Complete Worksheet • Pages 41-42 • (TG)			
	Day 13	Read Special Feature: Redwoods • Page 41 • (GDL)			
	Day 14	Read Lesson 8: Seeds • Pages 42-44 • (GDL) Complete Worksheet • Pages 43-45 • (TG)			
	Day 15				
Week 4	Day 16	Read Lesson 9: Monocots & Dicots • Pages 45-47 • (GDL) Complete Worksheet • Pages 47-48 • (TG)			
	Day 17	Read Lesson 10: Seeds—Where Are They? • Pages 48-51 • (GDL) Complete Worksheet • Pages 49-52 • (TG)			
	Day 18	Read Special Feature: George Washington Carver Pages 52-53 • (GDL)			
	Day 19	Complete **Flowering Plants & Seeds Quiz 2** (Lessons 5-10) Pages 311-312 • (TG)			
	Day 20				
Week 5	Day 21	Plants Unit 3: Roots & Stems Read Lesson 11: Roots • Pages 55-57 • (GDL) Complete Worksheet • Pages 53-54 • (TG)			
	Day 22	Read Lesson 12: Special Roots • Pages 58-60 • (GDL) Complete Worksheet • Pages 55-56 • (TG)			
	Day 23	Read Lesson 13: Stems • Pages 61-63 • (GDL) Complete Worksheet • Pages 57-58 • (TG)			
	Day 24	Read Lesson 14: Stem Structure • Pages 64-65 • (GDL) Complete Worksheet • Pages 59-60 • (TG)			
	Day 25				

Date	Day	Assignment	Due Date	✓	Grade
Week 6	Day 26	Read Lesson 15: Stem Growth • Pages 66-68 • (GDL) Complete Worksheet • Pages 61-62 • (TG)			
	Day 27	Complete **Roots & Stems Quiz 3** (Lessons 11-15) Pages 313-314 • (TG)			
	Day 28	Plants Unit 4: Leaves Read Lesson 16: Photosynthesis • Pages 70-73 • (GDL) Complete Worksheet • Pages 63-67 • (TG)			
	Day 29	Read Lesson 17: Arrangement of Leaves • Pages 74-76 • (GDL) Complete Worksheet • Pages 69-70 • (TG)			
	Day 30				
Week 7	Day 31	Read Lesson 18: Leaves—Shape & Design • Pages 77-80 • (GDL) Complete Worksheet • Pages 71-72 • (TG)			
	Day 32	Read Lesson 19: Changing Colors • Pages 81-83 • (GDL) Complete Worksheet • Pages 73-74 • (TG)			
	Day 33	Read Lesson 20: Tree Identification: Final Project Pages 84-86 • (GDL) Complete Worksheet • Pages 75-76 • (TG)			
	Day 34	Complete **Leaves Quiz 4** (Lessons 16-20) • Pages 315-316 • (TG)			
	Day 35				
Week 8	Day 36	Plants Unit 5: Flowers & Fruits Read Lesson 21: Flowers • Pages 88-90 • (GDL) Complete Worksheet • Pages 77-79 • (TG)			
	Day 37	Read Lesson 22: Pollination • Pages 91-93 • (GDL) Complete Worksheet • Pages 81-83 • (TG)			
	Day 38	Read Special Feature: Pierre-Joseph Redoute • Page 94 • (GDL)			
	Day 39	Read Lesson 23: Flower Dissection • Pages 95-97 • (GDL) Complete Worksheet • Pages 85-86 • (TG)			
	Day 40				
Week 9	Day 41	Read Special Feature: A Rose by Any Other Name • Page 98 • (GDL)			
	Day 42	Read Lesson 24: Fruits • Pages 99-101 • (GDL) Complete Worksheet • Pages 87-89 • (TG)			
	Day 43	Read Lesson 25: Annuals, Biennials, & Perennials Pages 102-104 • (GDL) Complete Worksheet • Pages 91-92 • (TG)			
	Day 44	Complete **Flowers & Fruits Quiz 5** (Lessons 21-25) Pages 317-318 • (TG)			
	Day 45				
First Semester-Second Quarter					
Week 1	Day 46	Plants Unit 6: Unusual Plants Read Lesson 26: Meat-eating Plants • Pages 106-108 • (GDL) Complete Worksheet • Pages 93-94 • (TG)			
	Day 47	Read Lesson 27: Parasites & Passengers • Pages 109-111 • (GDL) Complete Worksheet • Pages 95-96 • (TG)			
	Day 48	Read Lesson 28: Tropisms • Pages 112-114 • (GDL) Complete Worksheet • Pages 97-98 • (TG)			
	Day 49	Read Lesson 29: Survival Techniques • Pages 115-116 • (GDL) Complete Worksheet • Pages 99-100 • (TG)			
	Day 50				

Date	Day	Assignment	Due Date	✓	Grade
Week 2	Day 51	Read Lesson 30: Reproduction without Seeds Pages 117-119 • (GDL) • Complete Worksheet • Pages 101-102 • (TG)			
	Day 52	Read Lesson 31: Ferns • Pages 120-122 • (GDL) Complete Worksheet • Pages 103-104 • (TG)			
	Day 53	Read Lesson 32: Mosses • Pages 123-125 • (GDL) Complete Worksheet • Pages 105-106 • (TG)			
	Day 54	Read Lesson 33: Algae • Pages 126-128 • (GDL) Complete Worksheet • Pages 107-108 • (TG)			
	Day 55				
Week 3	Day 56	Read Lesson 34: Fungi • Pages 129-131 • (GDL) Complete Worksheet • Pages 109-110 • (TG)			
	Day 57	Complete **Unusual Plants Quiz 6** (Lessons 26-34) Pages 319-320 • (TG)			
	Day 58	Complete **World of Plants Final Exam** (Lessons 1-34) Pages 321-323 • (TG)			
	Day 59	Read Lesson 35: Conclusion • Page 132 • (GDL) Complete Worksheet • Page 111 • (TG)			
	Day 60				
Week 4	Day 61	The Human Body Unit 1: Body Overview Read Lesson 1: The Creation of Life • Pages 140-141 • (GDL) Complete Worksheet • Pages 115-116 • (TG)			
	Day 62	Read Lesson 2: Overview of the Human Body Pages 142-143 • (GDL) Complete Worksheet • Pages 117-121 • (TG)			
	Day 63	Read Special Feature: Leonardo da Vinci • Pages 144-145 • (GDL)			
	Day 64	Read Lesson 3: Cells, Tissues, & Organs • Pages 146-148 • (GDL) Complete Worksheet • Pages 123-125 • (TG)			
	Day 65				
Week 5	Day 66	Complete **Body Overview Quiz 1** (Lessons 1-3) Pages 327-328 • (TG)			
	Day 67	Body Unit 2: Bones & Muscles Read Lesson 4: The Skeletal System • Pages 150-152 • (GDL) Complete Worksheet • Pages 127-131 • (TG)			
	Day 68	Read Lesson 5: Names of Bones • Pages 153-155 • (GDL) Complete Worksheet • Pages 133-135 • (TG)			
	Day 69	Read Lesson 6: Types of Bones • Pages 156-158 • (GDL) Complete Worksheet • Pages 137-138 • (TG)			
	Day 70				
Week 6	Day 71	Read Lesson 7: Joints • Pages 159-161 • (GDL) Complete Worksheet • Pages 139-140 • (TG)			
	Day 72	Read Lesson 8: The Muscular System • Pages 162-164 • (GDL) Complete Worksheet • Pages 141-142 • (TG)			
	Day 73	Read Lesson 9: Different Types of Muscles • Pages 165-166 • (GDL) Complete Worksheet • Pages 143-144 • (TG)			
	Day 74	Read Lesson 10: Hands & Feet • Pages 167-169 • (GDL) Complete Worksheet • Pages 145-146 • (TG)			
	Day 75				

Date	Day	Assignment	Due Date	✓	Grade
Week 7	Day 76	Complete **Bones & Muscles Quiz 2** (Lessons 4-10) Pages 329-330 • (TG)			
	Day 77	Body Unit 3: Nerves & Senses Read Lesson 11: The Nervous System • Pages 171-173 • (GDL) Complete Worksheet • Pages 147-149 • (TG)			
	Day 78	Read Lesson 12: The Brain • Pages 174-176 • (GDL) Complete Worksheet • Pages 151-152 • (TG)			
	Day 79	Read Lesson 13: Learning & Thinking • Pages 177-179 • (GDL) Complete Worksheet • Pages 153-154 • (TG)			
	Day 80				
Week 8	Day 81	Read Special Feature: Brain Surgery • Pages 180-181 • (GDL)			
	Day 82	Read Lesson 14: Reflexes & Nerves • Pages 182-184 • (GDL) Complete Worksheet • Pages 155-156 • (TG)			
	Day 83	Read Lesson 15: The Five Senses • Pages 185-187 • (GDL) Complete Worksheet • Pages 157-158 • (TG)			
	Day 84	Read Lesson 16: The Eye • Pages 188-190 • (GDL) Complete Worksheet • Pages 159-160 • (TG)			
	Day 85				
Week 9	Day 86	Read Lesson 17: The Ear • Pages 191-193 • (GDL) Complete Worksheet • Pages 161-162 • (TG)			
	Day 87	Read Lesson 18: Taste & Smell • Pages 194-196 • (GDL) Complete Worksheet • Pages 163-164 • (TG)			
	Day 88	Complete **Nerves & Senses Quiz 3** (Lessons 11-18) Pages 331-332 • (TG)			
	Day 89	Body Unit 4: The Digestion System Read Lesson 19: The Digestive System • Pages 198-200 • (GDL) Complete Worksheet • Pages 165-168 • (TG)			
	Day 90				
		Mid-Term Grade			

Second Semester Suggested Daily Schedule

Date	Day	Assignment	Due Date	✓	Grade
		Second Semester-Third Quarter			
Week 1	Day 91	Read Lesson 20: Teeth • Pages 201-203 • (GDL) Complete Worksheet • Pages 169-170 • (TG)			
	Day 92	Read Lesson 21: Dental Health • Pages 204-205 • (GDL) Complete Worksheet • Pages 171-172 • (TG)			
	Day 93	Read Lesson 22: Nutrition • Pages 206-208 Complete Worksheet • Pages 173-174 • (TG)			
	Day 94	Read Special Feature: Florence Nightingale Pages 209-210 • (GDL)			
	Day 95				
Week 2	Day 96	Read Lesson 23: Vitamins & Minerals • Pages 211-213 • (GDL) Complete Worksheet • Pages 175-176 • (TG)			
	Day 97	Complete **Digestive System Quiz 4** (Lessons 19-23) Pages 333-334 • (TG)			
	Day 98	Body Unit 5: Heart & Lungs Read Lesson 24: The Circulatory System Pages 215-218 • (GDL) Complete Worksheet • Pages 177-178 • (TG)			
	Day 99	Read Lesson 25: The Heart • Pages 219-221 • (GDL) Complete Worksheet • Pages 179-181 • (TG)			
	Day 100				
Week 3	Day 101	Read Lesson 26: Blood • Pages 222-224 • (GDL) Complete Worksheet • Pages 183-186 • (TG)			
	Day 102	Read Special Feature: Blood — Who Needs It? Page 225 • (GDL)			
	Day 103	Read Lesson 27: The Respiratory System Pages 226-228 • (GDL) Complete Worksheet • Pages 187-188 • (TG)			
	Day 104	Read Lesson 28: The Lungs • Pages 229-231 • (GDL) Complete Worksheet • Pages 189-190 • (TG)			
	Day 105				
Week 4	Day 106	Complete **Heart & Lungs Quiz 5** (Lessons 24-28) Pages 335-336 • (TG)			
	Day 107	Body Unit 6: Skin & Immunity Read Lesson 29: The Skin • Pages 233-235 • (GDL) Complete Worksheet • Pages 191-192 • (TG)			
	Day 108	Read Lesson 30: Cross-section of Skin • Pages 236-238 • (GDL) Complete Worksheet • Pages 193-194 • (TG)			
	Day 109	Read Lesson 31: Fingerprints • Pages 239-242 • (GDL) Complete Worksheet • Pages 195-197 • (TG)			
	Day 110				

Date	Day	Assignment	Due Date	✓	Grade
Week 5	Day 111	Read Lesson 32: The Immune System • Pages 243-245 • (GDL) Complete Worksheet • Pages 199-200 • (TG)			
	Day 112	Read Lesson 33: Genetics • Pages 246-248 • (GDL) Complete Worksheet • Pages 201-203 • (TG)			
	Day 113	Read Special Feature: Gregor Mendel • Pages 249-250 • (GDL)			
	Day 114	Complete **Skin & Immunity Quiz 6** (Lessons 29-33) Pages 337-338 • (TG)			
	Day 115				
Week 6	Day 116	Read Lesson 34: Body Poster: Final Project Pages 251-252 • (GDL) Complete Worksheet • Pages 205-206 • (TG)			
	Day 117	Complete **The Human Body Final Exam** (Lessons 1-34) Pages 339-342 • (TG)			
	Day 118	Read Lesson 35: Conclusion • Page 253 • (GDL) Complete Worksheet • Page 207 • (TG)			
	Day 119	The World of Animals Unit 1: Mammals Read Lesson 1: The World of Animals • Pages 262-263 • (GDL) Complete Worksheet • Pages 211-212 • (TG)			
	Day 120				
Week 7	Day 121	Read Lesson 2: Vertebrates • Pages 264-265 • (GDL) Complete Worksheet • Pages 213-214 • (TG)			
	Day 122	Read Lesson 3: Mammals • Pages 266-268 • (GDL) Complete Worksheet • Pages 215-217 • (TG)			
	Day 123	Read Lesson 4: Mammals Large & Small Pages 269-272 • (GDL) Complete Worksheet • Pages 219-220 • (TG)			
	Day 124	Read Lesson 5: Monkeys & Apes • Pages 273-275 • (GDL) Complete Worksheet • Pages 221-223 • (TG)			
	Day 125				
Week 8	Day 126	Read Special Feature: Man & Monkeys • Pages 276-277 • (GDL)			
	Day 127	Read Lesson 6: Aquatic Mammals • Pages 278-281 • (GDL) Complete Worksheet • Pages 225-226 • (TG)			
	Day 128	Read Lesson 7: Marsupials • Pages 282-284 • (GDL) Complete Worksheet • Pages 227-228 • (TG)			
	Day 129	Complete **Mammals Quiz 1** (Lessons 1-7) Pages 345-346 • (TG)			
	Day 130				
Week 9	Day 131	Animals Unit 2: Birds & Fish Read Lesson 8: Birds • Pages 286-289 • (GDL) Complete Worksheet • Pages 229-232 • (TG)			
	Day 132	Read Special Feature: Charles Darwin • Page 290 • (GDL)			
	Day 133	Read Lesson 9: Flight • Pages 291-294 • (GDL) Complete Worksheet • Pages 233-235 • (TG)			
	Day 134	Read Lesson 10: The Bird's Digestive System Pages 295-297 • (GDL) Complete Worksheet • Pages 237-239 • (TG)			
	Day 135				

Date	Day	Assignment	Due Date	✓	Grade
		Second Semester-Fourth Quarter			
Week 1	Day 136	Read Lesson 11: Fish • Pages 298-300 • (GDL) Complete Worksheet • Pages 241-244 • (TG)			
	Day 137	Read Lesson 12: Fins & Other Fish Anatomy Pages 301-303 • (GDL) Complete Worksheet • Pages 245-247 • (TG)			
	Day 138	Read Lesson 13: Cartilaginous Fish • Pages 304-306 • (GDL) Complete Worksheet • Pages 249-250 • (TG)			
	Day 139	Complete **Birds & Fish Quiz 2** (Lessons 8-13) Pages 347-348 • (TG)			
	Day 140				
Week 2	Day 141	Animals Unit 3: Amphibians & Reptiles Read Lesson 14: Amphibians • Pages 308-310 • (GDL) Complete Worksheet • Pages 251-252 • (TG)			
	Day 142	Read Lesson 15: Amphibian Metamorphosis Pages 311-313 • (GDL) Complete Worksheet • Pages 253-254 • (TG)			
	Day 143	Read Lesson 16: Reptiles • Pages 314-316 • (GDL) Complete Worksheet • Pages 257-258 • (TG)			
	Day 144	Read Special Feature: When Did the Dinosaurs Live? Pages 317-318 • (GDL)			
	Day 145				
Week 3	Day 146	Read Lesson 17: Snakes • Pages 319-321 • (GDL) Complete Worksheet • Pages 259-260 • (TG)			
	Day 147	Read Special Feature: Rattlesnakes • Page 322 • (GDL)			
	Day 148	Read Lesson 18: Lizards • Pages 323-325 • (GDL) Complete Worksheet • Pages 261-262 • (TG)			
	Day 149	Read Lesson 19: Turtles & Crocodiles • Pages 326-328 • (GDL) Complete Worksheet • Pages 263-265 • (TG)			
	Day 150				
Week 4	Day 151	Complete **Amphibians & Reptiles Quiz 3** (Lessons 14-19) Pages 349-350 • (TG)			
	Day 152	Animals Unit 4: Arthropods Read Lesson 20: Invertebrates • Pages 330-332 • (GDL) Complete Worksheet • Pages 267-268 • (TG)			
	Day 153	Read Lesson 21: Arthropods • Pages 333-335 • (GDL) Complete Worksheet • Pages 269-271 • (TG)			
	Day 154	Read Lesson 22: Insects • Pages 336-338 • (GDL) Complete Worksheet • Pages 273-275 • (TG)			
	Day 155				

Date	Day	Assignment	Due Date	✓	Grade
Week 5	Day 156	Read Lesson 23: Insect Metamorphosis • Pages 339-341 • (GDL) Complete Worksheet • Pages 277-279 • (TG)			
	Day 157	Read Lesson 24: Arachnids • Pages 342-344 • (GDL) Complete Worksheet • Pages 281-282 • (TG)			
	Day 158	Read Lesson 25: Crustaceans • Pages 345-346 • (GDL) Complete Worksheet • Pages 283-284 • (TG)			
	Day 159	Read Lesson 26: Myriapods • Pages 347-349 • (GDL) Complete Worksheet • Pages 285-286 • (TG)			
	Day 160				
Week 6	Day 161	Complete **Arthropods Quiz 4** (Lessons 20-26) Pages 351-352 • (TG)			
	Day 162	Animals Unit 5: Other Invertebrates Read Lesson 27: Mollusks • Pages 351-353 • (GDL) Complete Worksheet • Pages 287-288 • (TG)			
	Day 163	Read Lesson 28: Cnidarians • Pages 354-357 • (GDL) Complete Worksheet • Pages 289-291 • (TG)			
	Day 164	Read Lesson 29: Echinoderms • Pages 358-360 • (GDL) Complete Worksheet • Pages 293-294 • (TG)			
	Day 165				
Week 7	Day 166	Read Lesson 30: Sponges • Pages 361-362 • (GDL) Complete Worksheet • Pages 295-296 • (TG)			
	Day 167	Read Lesson 31: Worms • Pages 363-365 • (GDL) Complete Worksheet • Pages 297-298 • (TG)			
	Day 168	Complete **Other Invertebrates Quiz 5** (Lessons 27-31) Pages 353-354 • (TG)			
	Day 169	Animals Unit 6: Simple Organisms Read Lesson 32: Kingdom Protista • Pages 367-369 • (GDL) Complete Worksheet • Pages 299-300 • (TG)			
	Day 170				
Week 8	Day 171	Read Lesson 33: Kingdom Monera & Viruses Pages 370-372 • (GDL) Complete Worksheet • Pages 301-302 • (TG)			
	Day 172	Read Special Feature: Louis Pasteur — Got Milk? Pages 373-374 • (GDL)			
	Day 173	Complete **Simple Organisms Quiz 6** (Lessons 32-33) Pages 355-356 • (TG)			
	Day 174	Read Lesson 34: Animal Notebook: Final Project Page 375-376 Complete Worksheet • Pages 303-304 • (TG)			
	Day 175				
Week 9	Day 176	Study day for final exam			
	Day 177	Study day for final exam			
	Day 178	Complete **World of Animals Final Exam** (Lessons 1-34) Pages 357-360 • (TG)			
	Day 179	Read Lesson 35: Conclusion • Page 377 • (GDL) Complete Worksheet • Page 305 • (TG)			
	Day 180				
		Final Grade			

Plant Worksheets

for Use with

The World of Plants

(*God's Design: Life* Series)

Name _____   Date _____

🏅 Grass Comparison Worksheet

Type of seed	Observations of seed	Date planted	Date first leaves observed	Observations after 1 week	Observations after 2 weeks

Which seeds germinated first? _____

How are the leaves similar? _____

How are the leaves different? _____

Which plant is the tallest? _____

Which is the shortest? _____

The World of Plants

| | *God's Design: Life* | The World of Plants | Day 12 | Unit 2 Lesson 7 | Name |

7 Trees

Did George Washington really chop down the cherry tree?

Supply list

☐ Index cards labeled with vocabulary words

☐ Markers or crayons

What Kind of Tree Is This?

Use the instructions on page 39 of your student book to learn the differences between deciduous and evergreen trees. Be sure to draw the images on the back of the index cards as instructed. It will help you successfully complete the activity!
What were your results?

Supplies for Challenge

☐ Drawing materials

Tree Shapes Challenge

This challenge helps you lean about the growth habit of different trees. On page 40 of your student book, you will find detailed instructions to follow. Please try to do as least three or four different types of trees to get a better understanding of why there are differences among their growth.
Note: When you complete this activity, be sure to save the drawings to be part of your final project.

What did we learn?

1. What makes a plant a tree?

2. How are deciduous and evergreen trees different?

3. How are angiosperms and gymnosperms different?

🚀 Taking it further

1. Do evergreen trees have growth rings?

2. How long do you think a tree lives?

| | God's Design: Life | The World of Plants | Day 16 | Unit 2 Lesson 9 | Name |

9 Monocots & Dicots

What's inside that seed?

 Supply list

☐ Several bean seeds (pinto, kidney, etc.) **Note:** Soak these seeds in water for approx. 24 hours prior to use.

☐ Several corn seeds (not popcorn) **Note:** Soak these seeds in water for approx. 24 hours prior to use.

☐ Dissecting scalpel or very sharp knife (for adult use only)

☐ Magnifying glass

☐ Jar or plastic cup

☐ Paper towels

Seed Dissection Activity

Complete the activity on page 46 and then answer the following questions. Be sure to do both procedures and allow ample time for the student to make their observations.

1. How has the water affected the seed coat?

2. How does the seed coat of the corn differ from the seed coat of the beans?

Supplies for Challenge

☐ Other seeds for dissection

Where do seeds germinate?

Read this text on page 47 of your student book. Explain to your instructor what the difference is between hypogeal and epigeal germination. Which of the plants (bean and corn) experience hypogeal germination and which experience epigeal germination?

What did we learn?

1. What differences did you observe between the monocot and dicot seeds?

2. What parts of each seed were you able to identify?

3. What is the plumule?

4. What is the radicle?

5. What is the purpose of the cotyledon?

Taking it further

1. Why did you need to soak the seeds before dissecting them?

2. What differences do you think you might find in plants that grow from monocot and dicot seeds?

God's Design: Life | The World of Plants | Day 17 | Unit 2 Lesson 10 | Name

Seeds—Where Are They?

How do they get around?

🧪 Supply list

☐ Several different fruits and vegetables (apple, tomato, peach, etc.)

☐ Baking dish lined with aluminum foil

☐ Several pinecones with scales tightly shut

☐ Copy of "Seeds Get Around" worksheet

🧪 Seed Location Activity

Instructions are found on page 50 of your student book.

1. Which fruits did you use?

2. Which seeds have softer seed coats? Which are harder?

3. Which seeds appear to be dicots? Which are monocots?

4. In observing the pinecones, are there any seeds easily seen?

🧪 Seed Dispersal Activity

Also, on page 50 of your student book, there is an activity to learn how seeds are dispersed in the environment. Be sure to complete the Seeds Get Around Worksheet on page 51 of this teacher guide.

🎖 Supplies for Challenge

☐ Several different kinds of seeds

☐ Whole coconut or coconut seed, if available

☐ Copy of "Water Dispersal Test"

The World of Plants 49

Water Dispersal Challenge

Isn't it interesting the different ways seeds are dispersed? Complete the Water Dispersal Test Worksheet on page 52 of this teacher guide using the instructions found for this lesson in your student book. Why is seed dispersal important to plant survival?

What did we learn?

1. What are three ways seeds can be moved or dispersed?
   a.

   b.

   c.

2. Where are good places to look for seeds?

Taking it further

1. How do people aid in the dispersal of seeds?

2. What has man done to change or improve seeds or plants?

3. If a seed is small, will the mature plant also be small?

4. Do the largest plants always have the largest seeds?

5. Why do you think God created many large plants to have small seeds?

6. Can you name a plant that disperses its seeds by the whole plant blowing around?

Name _____ Date _____

🧪 Seeds Get Around Worksheet

By Wind

Draw a picture of a plant with seeds dispersed by wind. If possible, glue a seed below.

What has man designed that works in a similar way?

By Animals

Draw a picture of a plant with seeds dispersed by animals. If possible, glue a seed below.

What has man designed that works in a similar way?

By Explosions

Draw a picture of a plant with seeds dispersed by explosions. If possible, glue a seed below.

What has man designed that works in a similar way?

Name _____ Date _____

Water Dispersal Test Worksheet

Type of seed	Hypothesis: Will it float?	Results: Did it float?

| God's Design: Life | The World of Plants | Day 22 | Unit 3 Lesson 12 | Name |

12 Special Roots

Not always underground

Supply list

☐ Onion with roots

☐ Flower bulbs (tulips, daffodils, etc.), if available

Adventitious and Aerial Root Observations

Look at the onion. Look at the roots. Describe what you see. Now look at the tulip or daffodil bulbs. Tell your teacher how the bulbs are different or similar.

Root Poster

Follow the directions in your student book on page 60. Be sure to draw or collect photos of different roots from a magazine or other sources. Be sure to include the various types of roots.

Supplies for Challenge

☐ Poster board

☐ Pictures of special roots

☐ Drawing materials

Banyan Tree Challenge

1. What is another name for the banyan tree?

2. How is it an epiphyte?

3. Banyan trees are native to what country?

4. Name two uses for the banyan tree.

    a.

    b.

What did we learn?

1. What are adventitious roots?

2. What are aerial roots?

3. What are prop roots?

Taking it further

1. Why do you think that some plants have specialized roots?

2. Why do some plants need prop roots?

| *God's Design: Life* | The World of Plants | Day 23 | Unit 3 Lesson 13 | Name

13 Stems

Connecting it all together

🧪 Supply list

☐ Stalk of celery with leaves

☐ Glass of water

☐ Food coloring (red or blue)

☐ 2 clear plastic cups with potting soil

☐ Jars with beans and corn from lessons 8 and 9

🧪 What Goes Up? Activity

Follow the procedure listed on page 62 of your student book. Be sure to tell your teacher about what you observe. Make sure to compare the stems of the corn and bean plants you are growing. Make note of any similarities or differences.

🏅 Water Movements in Plants Challenge

What are the three different processes involved in moving water and nutrients up a plant?

a.

b.

c.

Draw an image showing these processes at work in a plant — be creative in showing how they function.

What did we learn?

1. What are the main functions of a stem?

2. What do we call the stem of a tree?

Taking it further

1. If a tree branch is 3 feet above the ground on a certain day, how far up will the branch be 10 years later?

2. What are some stems that are good to eat?

| | God's Design: Life | The World of Plants | Day 24 | Unit 3 Lesson 14 | Name |

14 Stem Structure

How they are put together

🧪 Supply list

☐ Drawing materials

☐ Plant with new growth, if available

🧪 Examining Stems Activity

Read the instructions on page 65 of your textbook. You will need to be able to see new growth on a bush or tree to complete this activity. **Note**: If you cannot find a bush or tree to observe new growth, consider visiting a nursery or looking at seasonal plants available at retail stores throughout the year.

🏅 Branching Challenge

Take a second look at the plants you just examined and tell your teacher the answer to the following questions:

1. Is the growth primarily vertical or horizontal?

2. Can you easily see a trunk or primary stem most of the way up the plant?

3. Which plants in your yard have excurrent branching and which have deliquescent branching?

🧠 What did we learn?

1. What are the major structures of a stem?

2. Where does new growth occur on a stem?

3. What gives the plant its size and shape?

🚀 Taking it further

1. What will happen to a plant if its terminal buds are removed?

2. How are stems different between trees and bushes?

3. In your experience, do flower stems have the same structures, including terminal buds, nodes, etc., as bush and tree stems?

| | God's Design: Life | The World of Plants | Day 26 | Unit 3 Lesson 15 | Name |

15 Stem Growth

Farther up and farther out

Supply list (No supplies needed)

Looking at tree rings

Look at the picture of tree rings on page 67 of your student book. Try to answer the following questions:

1. Do you notice any rings that are significantly wider or narrower than the rest?

2. What can you guess about growing conditions when the rings are wider or narrower?

3. Can you determine by the rings how old the tree is?

Challenge: Vascular Tissue

In lesson 13, you watched fluids moving up the stem of a stalk of celery. Do you remember how the xylem were arranged in the celery? They were arranged in a circular pattern. Would that indicate that celery is a monocot or a dicot?

What did we learn?

1. What are epidermis cells?

2. What is bark?

3. Name three types of cells inside a stem.

🚀 Taking it further

1. Can we tell a tree's age from the rings inside the trunk? Why or why not?

2. If you wanted to make a very strong wooden spoon, which part of the tree would you use?

3. Why don't herbaceous plants have bark?

| God's Design: Life | The World of Plants | Day 28 | Unit 4 Lesson 16 | Name |

16 Photosynthesis

Making food for the world

🧪 Supply list

☐ 3 mint plants or other fast-growing plants
☐ Copy of "Photosynthesis Data Sheet"
☐ Scissors
☐ 2 cardboard boxes that are big enough to cover a plant and allow for growth
☐ Liquid measuring cup

🧪 Sunlight and Photosynthesis Activity

Let's test the effects of sunlight on photosynthesis. Follow the steps on page 72 in your student book and complete the Photosynthesis Data Sheet on page 67 of this teacher guide.

🏅 Supplies for Challenge

☐ Copy of "Photosynthesis Building Blocks" worksheet
☐ Scissors
☐ Tape

🏅 The Photosynthesis Reaction Challenge

On page 73 in your student book you will find instructions for how to complete this challenge. The Photosynthesis Building Blocks are on page 65 of this teacher guide. Did this help you to understand how photosynthesis works on a molecular level? Explain the process to your teacher using your project as an example.

🧠 What did we learn?

1. What are the "ingredients" needed for photosynthesis?

2. What are the "products" of photosynthesis?

3. How did God specifically design plants to be a source of food?

4. How does carbon dioxide enter a leaf?

🚀 Taking it further

1. On which day of creation did God create plants?

2. On which day did He create the sun?

3. In our experiment, we found that the plant that got less sunlight grew more slowly than the one that had full sunlight. Is this true for all plants?

Name _____ Date _____

🧪 Photosynthesis Data Sheet

Hypothesis: Which plant do you think will grow the fastest? _____

Date	Plant A height (Full sun)	Plant B height (Part sun)	Plant C height (No sun)	Observations

Conclusion

What did you learn about how sunlight affects plant growth?

_____

_____

_____

The World of Plants

Name _____ Date _____

🎖 Photosynthesis Building Blocks

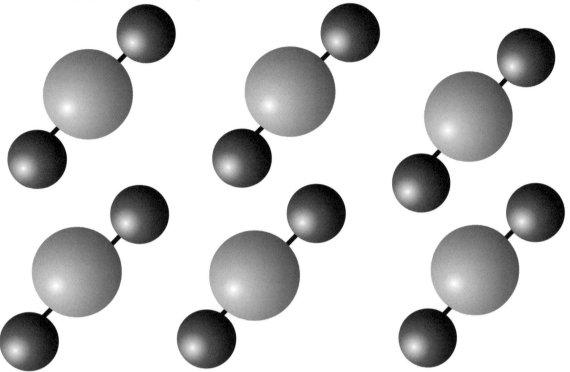

Carbon dioxide

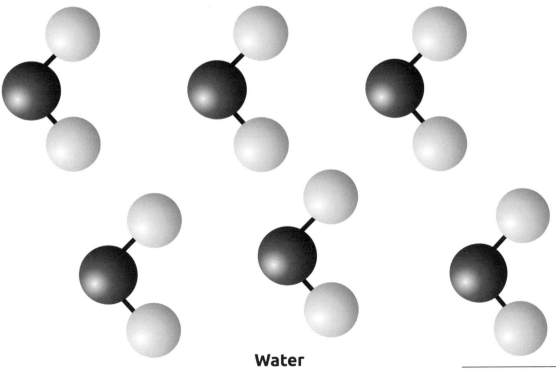

Water

The World of Plants 67

17 Arrangement of Leaves

Maximizing sunlight

🧪 Supply list

☐ Drawing paper

☐ Crayons, markers, or colored pencils

🧪 Observing Leaf Arrangement

For this activity, follow the instructions found on page 75 of your student book. Be sure to do enough to create a small book with a cover you make so you can show others what you have learned.

🏅 Supplies for Challenge

☐ Aloe plant or other succulent, if available

🏅 Special Leaves Challenge

What are the three special leaves that you learned about in this challenge?
See how many plants or photos of plants you can find with these special leaves.

🧠 What did we learn?

1. What are four common ways leaves can be arranged on a plant?

    a.

    b.

    c.

    d.

2. Why do you think God created each of these different leaf arrangements?

3. Why is it important for sunlight to reach each leaf?

🚀 Taking it further

1. How does efficient leaf arrangement show God's provision or care for us?

2. What other feature, besides leaf arrangement, aids leaves in obtaining maximum exposure to sunlight?

| God's Design: Life | The World of Plants | Day 31 | Unit 4 Lesson 18 | Name |

18 Leaves—Shape & Design

What's your shape?

🧪 Supply list

☐ 1 or more large leaves freshly picked from a tree
☐ Knife or scissors
☐ Paper and colored pencils
☐ Red food coloring
☐ Grass
☐ Bean and corn plants from lessons 8 and 9

🧪 Observing Leaf Shapes and Vein Arrangements

Complete this activity on page 79 of the student book and answer the questions below.

1. How are the shapes and vein arrangements different?

2. Which plant has broad leaves?

3. Which has long narrow leaves?

4. Which plant is a monocot?

5. Which plant is a dicot?

Leaf Shapes & Margins Challenge

Fill in the blank.

1. The edge of a leaf is called _____.

2. A leaf with an entire margin has _____ edges.

3. Leaves with a jagged edge have a _____ margin.

4. Leaves with large indentions on the edges are said to have _____ margins.

5. If a petiole has several leaflets growing from it, then it is called a _____ leaf.

What did we learn?

1. What general shape of leaves do monocots and dicots have?

2. How can we use leaves to help us identify plants?

3. How do nutrients and food get into and out of the leaves?

Taking it further

1. Describe how the arrangement of the veins is most efficient for each leaf shape.

| God's Design: Life | The World of Plants | Day 32 | Unit 4 Lesson 19 | Name |

19 Changing Colors

The beauty of autumn

🧪 Supply list

☐ If it is autumn, collect different colored leaves. If not, use several colors of construction paper
☐ Scissors
☐ Glue
☐ Tagboard/poster board
☐ Newspaper
☐ Heavy book

🧪 An Autumn Picture Activity

Follow the instructions on page 82 of your student book to make a picture out of autumn colors.

🏅 Supplies for Challenge

☐ 2 or 3 different fresh leaves
☐ Fingernail polish remover
☐ Coffee filters
☐ Quarter or other coin
☐ Tape
☐ Dish

🏅 Leaf Pigments Challenge

Use paper chromatography to determine what pigments are present in a plant by following the instructions on page 82 of your student book.
Tell your teacher what you found during the challenge.

1. Did each leaf have the color of pigments you expected?

The World of Plants ⧸⧸ 73

2. Are you surprised to find so many different colors in them?

What did we learn?

1. How do trees know when to change color?

2. Why do trees drop their leaves?

3. Why don't evergreen trees drop their leaves in the winter?

Taking it further

1. Do trees and bushes with leaves that are purple in the summer still have chlorophyll?

2. What factors, other than daylight, might affect when a tree's leaves start changing color?

God's Design: Life | The World of Plants | Day 33 | Unit 4 Lesson 20 | Name

Tree Identification

How do I know what tree it is?

🧪 Final Project supply list

☐ Tree field guide such as *Peterson's First Guide to Trees, Reader's Digest Field Guide of North America,* or *Trees of North America* by C. Frank Brockman

☐ Zipper bags or other storage containers

☐ Leaf press or heavy books and newspaper

☐ Access to trees with leaves

☐ Index cards

☐ Colored pencils

☐ Colored paper

☐ Photo album with magnetic pages or 3-ring binder with plastic sheet protectors

🧪 Final Project — Leaf Notebook

Note: This three-part activity will take several days to complete using the instructions on pages 85–86 of your student book.

🏅 Supplies for Challenge

☐ Growth habit drawings from lesson 7

🏅 Finishing Your Notebook

This is a chance to get creative! Include information on the growth habits you learned in lesson 7 of this course, and see if you can make your leaf notebooks something very interesting to share with others.

🧠 What did we learn?

1. What are some ways you can try to identify a plant?

2. What are the biggest differences between deciduous and coniferous trees?

🚀 Taking it further

1. Why do we need to be able to identify trees and other plants?

God's Design: Life | The World of Plants | Day 36 | Unit 5 Lesson 21 | Name

Flowers
The beauty of sight and scent

🧪 Supply list

☐ Copy of "Flower Pattern"
☐ Green pipe cleaners
☐ Flexible soda straws
☐ Modeling clay
☐ Colored construction paper

☐ Scissors
☐ Glue
☐ Hole punch
☐ Corn meal or yellow sand

🧪 Flower Model Activity

Directions for this activity are on page 89 of the student book. The flower pattern sheet is on page 79 of this teacher guide.

🏅 Composite Flower Challenge

There are a lot of different kinds of flowers! See if you can describe each of the following types of flowers to your teacher:

1. composite flowers

2. disk flowers

3. ray flowers

The World of Plants // 77

What did we learn?

1. What are the four parts of the flower and what is the purpose or job of each part?

   a.

   b.

   c.

   d.

Taking it further

1. Why do you think God made so many different shapes and colors of flowers?

Name _____ Date _____

🧪 Flower Pattern

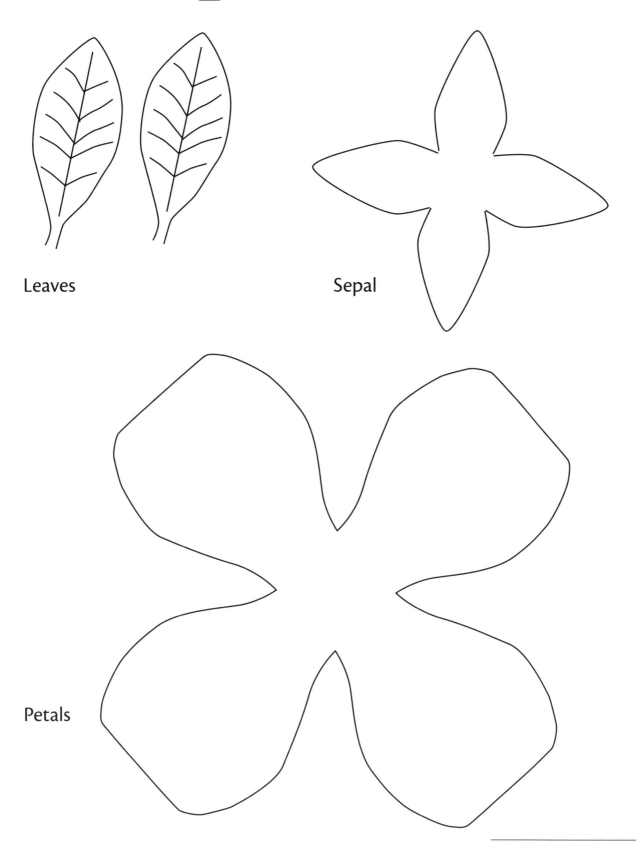

Leaves

Sepal

Petals

The World of Plants **//** 79

God's Design: Life | The World of Plants | Day 37 | Unit 5 Lesson 22 | Name

Pollination

The buzzing bee's job

🧪 Supply list

☐ Copy of the "Flip Book" worksheet

☐ Crayons, markers, or colored pencils

☐ Stapler

☐ Bean and corn plants from lessons 8 and 9

🧪 Pollination Flip Book Activity

Create your own book to illustrate pollination. Using the Flip Book Worksheet on page 83 of this teacher guide, follow the instructions on page 92 of your student book.

🎖 Supplies for Challenge

☐ Microscope and slide, if available

☐ Pollen grains

☐ Magnifying glass

🎖 Pollen Challenge

After reading the text on pages 92-93 in your student book, collect some samples of pollen from various flowers. Look at them under a microscope or with a magnifying glass.

Write a short report on or do a short oral report of what you see.

The World of Plants 81

What did we learn?

1. What animals can pollinate a flower?

2. How can a flower be pollinated without an animal?

3. Does pollen have to come from another flower?

Taking it further

1. Why do you suppose God designed most plants to need cross-pollination?

Name _____   Date _____

🧪 Flip Book

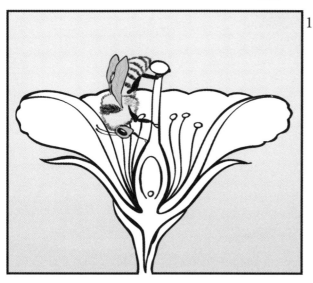

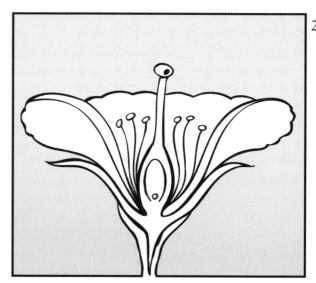

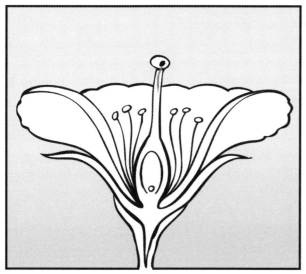

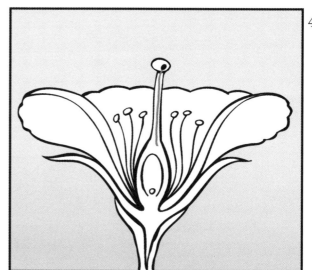

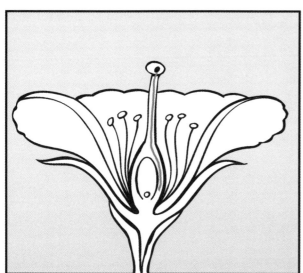

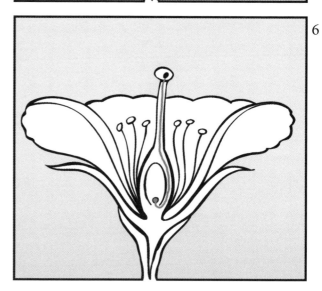

The World of Plants **83**

| God's Design: Life | The World of Plants | Day 39 | Unit 5 Lesson 23 | Name |

23 Flower Dissection

Seeing what's inside

🧪 Supply list

☐ A fresh flower with easily visible reproductive parts (a lily or an alstroemeria is a good example)

☐ Sharp knife or utility knife

🧪 Flower Dissection

You are getting ready to dissect a flower and learn more about its different parts. Follow the instructions on page 96 of your student book. Caution! Using the sharp knife or utility knife requires supervision by your teacher. Be very careful to keep your hands and fingers out of the path of the blade.

Note: The Optional Activity noted on page 96 of the student book can be considered a bonus activity for additional points if needed.

🏅 Supplies for Challenge

☐ Composite flower such as a daisy, sunflower, or zinnia

🏅 Composite Flower Dissection

The procedure on page 97 of the student book will allow you to dissect a compound flower and examine its parts. After completing the challenge, make sure you can answer the following questions for your teacher.

1. How do the flowers in the composite flower compare to the flower you previously dissected?

2. How are they the same?

3. How are they different?

What did we learn?

1. How many ovules did you find?

2. What did they look like? If possible, compare them to the mature seeds that are ready to be planted.

Taking it further

1. Why are the ovules in the flower green or white when most seeds are brown or black?

2. If you planted the ovules, would they grow into a plant?

| God's Design: Life | The World of Plants | Day 42 | Unit 5 Lesson 24 | Name |

24 Fruits

Is it ripe yet?

🧪 Supply list

☐ Apple, strawberry, pineapple (all fresh and whole, if possible)
☐ Knife

🧪 Fruit Identification

Learn how different fruits form by examining the fruits in the activity on page 100 of your student book.

🏅 Supplies for Challenge

☐ Copy of "Fruit Classification" worksheet

🏅 Fruit Divisions

Read about Fruit Divisions on page 101 of the student book, then complete the Fruit Classification Worksheet on page 89 of this teacher guide.

🧠 What did we learn?

1. What is the main purpose of fruit?

2. What are the three main groups of fruit?

3. Describe how each type of fruit forms.

The World of Plants // 87

Taking it further

1. What is the fruit of a wheat plant?

2. Which category of fruit is most common?

3. Why do biologists consider a green pepper to be a fruit?

Name _____   Date_____

🎖 Fruit Classification Worksheet

Next to each fruit listed below, write the letter indicating which type of fruit it is.

1. _____ Acorn   A. Drupe

2. _____ Pea   B. Berry

3. _____ Pear   C. Pome

4. _____ Avocado   D. Nut

5. _____ Mango   E. Legume

6. _____ Peanut   F. Grain

7. _____ Lima bean

8. _____ Wheat

9. _____ Nectarine

10. _____ Green pepper

11. _____ Pecan

12. _____ Crabapple

13. _____ Grapefruit

14. _____ Corn

15. _____ Rice

The World of Plants

| God's Design: Life | The World of Plants | Day 43 | Unit 5 Lesson 25 | Name |

25 Annuals, Biennials, & Perennials

How long do they grow?

🧪 Supply list

☐ Copy of "Plant Word Search"

🧪 Plant Word Search

Complete Word Search on next page of this teacher guide.

🏅 Ephemerals Challenge

Give a short presentation about ephemeral plants — where they are found, why they have short lifecycles, and why some have a connection to Genesis 3:17-19.

🧠 What did we learn?

1. What is an annual plant?

2. What is a biennial plant?

3. What is a perennial plant?

🚀 Taking it further

1. Why don't we often see the flowers of biennial plants?

2. Why don't people grow new plants from the seeds produced by the annuals each year?

The World of Plants 91

Name _____   Date _____

🧪 Plant Word Search

Find the following words in the puzzle below.

Annual	Roots	Seeds	Chlorophyll
Biennial	Stem	Photosynthesis	Pinnate
Perennial	Leaves	Pollination	Palmate
Fruit	Flowers	Cotyledon	

```
P H O T O S Y N T H E S I S S
H R S E E D S K Q E R T A L P
O G T E E S A A N N U A L V A
S T E R B T M C H R Y S E E S
T B M P H O R O S Y N I A C H
C H L O R O P H Y L L M V O P
Z O P O N P U P U W R F E T E
L H D R M A B I P R U L S Y R
V I P O L L I N A T I O N L E
O R A O C M E N T L A W Q E N
P I N T E A N A I P T E C D N
A B C S E T N T B A I R O O I
L E V B W E I E C C A S T N A
M A N V C P A R E S N I A L L
A F R U I T L D E M F R I U R
```

92 ∥ The World of Plants

| God's Design: Life | The World of Plants | Day 46 | Unit 6 Lesson 26 | Name |

26 Meat-eating Plants

Will it eat me?

Supply list

☐ Small box

☐ Stick or pole

Making a Trap

See if you can make the simple trap described on page 107 of your student book. How is it similar or different than the unusual plant traps you learned about today?

Supplies for Challenge

☐ Drawing materials

Cobra Lily Challenge

The cobra lily is an interesting plant with a trap that insects cannot see. If you were to design a carnivorous plant what would it look like? How would it trap insects or other animals? Use your imagination and draw a picture of your meat-eating plant. Be sure to give it a name.

What did we learn?

1. What is a carnivorous plant?

2. Why do some plants need to be carnivorous?

3. How does a carnivorous plant eat an insect?

🚀 Taking it further

1. Where are you likely to find carnivorous plants?

2. How might a Venus flytrap tell the difference between an insect on its leaf and a raindrop?

| God's Design: Life | The World of Plants | Day 47 | Unit 6 Lesson 27 | Name |

27 Parasites & Passengers

Living off of each other

🧪 Supply list

☐ Drinking straw

☐ Field guide to plants

☐ Knife or scissors

☐ Sink

☐ Coffee stirrer (a narrow straw)

🧪 Parasite Model Activity

Discover how a parasite steals nutrients from its host by following the activity directions on page 110 of the student book. When you complete the activity, tell your teacher what you have learned. **Note:** The Search for Parasites & Passengers activity on page 110 can be considered an optional seasonal activity.

🏅 Supplies for Challenge

☐ Research materials—depends on your chosen topic

🏅 Plant Research Challenge

Follow the instructions on page 111 of the student book. Which plant did you choose? Be sure to draw a picture to go with your presentation.

🧠 What did we learn?

1. What is a parasitic plant?

2. What is a passenger plant?

3. How do passenger plants obtain water and minerals?

🚀 Taking it further

1. Where is the most likely place to find passenger plants?

2. Do passenger plants perform photosynthesis?

3. Do parasitic plants perform photosynthesis?

| God's Design: Life | The World of Plants | Day 48 | Unit 6 Lesson 28 | Name |

28 Tropisms

How plants respond

🧪 Supply list

☐ A houseplant with several leaves

🧪 Observing Heliotropism

Using a house plant, you will learn about heliotropism. Follow the directions on page 113 of your student book. Make sure you turn the plant at least twice to see how the leaves are affected.

🎖 Supplies for Challenge

☐ Copy of "Tropisms" worksheet

🎖 More Tropisms

Complete the Tropisms Worksheet on the next page by drawing a picture demonstrating each tropism listed.

🧠 What did we learn?

1. What is geotropism?

2. What is hydrotropism?

3. What is phototropism or heliotropism?

🚀 Taking it further

1. Why are tropisms sometimes called "survival techniques"?

2. Will a seed germinate if it is planted 5 feet (1.5 m) from the water?

3. Where are some places you would not want to plant water-seeking plants such as willows?

Name _____   Date _____

🏅 Tropisms Worksheet

Hydrotropism	Phototropism
Geotropism	Chemotropism
Thermotropism	Thigmotropism

| God's Design: Life | The World of Plants | Day 49 | Unit 6 Lesson 29 | Name

 **Survival Techniques**

Surviving in harsh climates

 Supply list

☐ Cactus plant
☐ Magnifying glass

Examining a Cactus

After completing this activity on page 116 of the student book, discuss how this plant was designed to survive the harsh climate of a desert.

Supplies for Challenge

☐ Copy of "Designed for Survival" worksheet

Designed for Survival

Complete the Designed for Survival Worksheet on the next page. You can review previous lessons to have information and ideas to also include on the worksheet.

What did we learn?

1. How do some plants survive in hot, dry climates?

2. How do some plants survive in cold, windy climates?

Taking it further

1. Why do alpine plants need protection from the sun?

The World of Plants ▰▰ 99

Name _____   Date _____

🎖 Designed for Survival Worksheet

List six things that plants need to survive.

_____   _____

_____   _____

_____   _____

List six things that can harm plants.

_____   _____

_____   _____

_____   _____

List 12 ways that plants have been designed to survive. Think about all the amazing things that plants regularly do and also think about some of the special designs that some plants have.

_____   _____

_____   _____

_____   _____

_____   _____

_____   _____

_____   _____

List four ways that people help plants to survive.

_____

_____

_____

_____

| God's Design: Life | The World of Plants | Day 51 | Unit 6 Lesson 30 | Name |

 Reproduction Without Seeds

There are other ways

🧪 Supply list

☐ Potato

☐ Jar

☐ Potting soil

☐ Water

🧪 Growing a New Potato Plant

Note: The activity on page 118 of the student book can be assembled in a few moments, but it will take 10 to 14 days to see a result.

🏅 Supplies for Challenge

☐ Research materials on genetic modification

🏅 Cloning Plants Challenge

Do some research to find out how genetic modification works, how it is being used, what foods you might be eating that are GMOs, and what controversies surround this interesting field of science. Be sure to share what you learn with your teacher.

🧠 What did we learn?

1. What are some ways that plants can reproduce without growing from seeds?

🚀 Taking it further

1. Why can a potato grow from a piece of potato instead of from a seed?

2. Will the new plant be just like the original plant?

 Ferns

Seedless plants

🧪 Supply list

☐ Paper
☐ Paint and paint brushes
☐ Glue
☐ Corn meal or yellow sand
☐ Fresh fern frond, if available

🧪 Fern Fronds

Using the image and instructions on page 121 of the student book, complete this activity using a drawing of a fern frond and cornmeal or sand to show the spores. Leave space on the drawing to be able to complete the Fern Structure Challenge if you are doing it as well.

🎖 Fern Structure Challenge

After you read more about the structure of ferns on pages 121-122, add ground, rhizomes, and roots to your Fern Fronds Activity painting to show the whole structure of a fern plant.

🧠 What did we learn?

1. How are ferns like other plants?

2. What are fern leaves called?

3. How are ferns different from other plants?

4. How do they reproduce?

🚀 Taking it further

1. Why can't ferns reproduce with seeds?

| God's Design: Life | The World of Plants | Day 53 | Unit 6 Lesson 32 | Name |

32 Mosses

Do you really find moss on the north side of trees?

🧪 Supply list

☐ Paper and glue

☐ Colored pencils or crayons

☐ Magnifying glass

☐ Dried moss from a craft shop

🧪 Find the Moss Activity

Complete the drawing assignment on page 124 of your student book and then try to find living moss that you can observe with a magnifying glass. Describe what you see.

🏅 Supplies for Challenge

☐ Peat moss

☐ Dirt or soil from your yard

☐ Paper cups

🏅 Peat Moss Challenge

Learn about the water-absorbing qualities of peat moss in this activity from page 125 in the student book. Following the challenge, be sure to share what you observed with your teacher.

🧠 What did we learn?

1. How do mosses differ from seed-bearing plants?

2. How do mosses differ from ferns?

3. How do mosses produce food?

🚀 Taking it further

1. Are you likely to find moss in a desert? Why/why not?

God's Design: Life | The World of Plants | Day 54 | Unit 6 Lesson 33 | Name

33 Algae

Are all green things plants?

🧪 Supply list

☐ Paper
☐ Colored pencils
☐ Glue
☐ Construction paper
☐ Scissors

🧪 Food Chain Picture

See page 127 in the student book for additional details and the procedure so you can make a food chain.

🏅 Supplies for Challenge

☐ Sample of pond water
☐ Microscope and slide

🏅 Amazing Algae Challenge

Be sure to draw the algae you see in the sample of pond water. For more details, see page 128 in the student book.

🧠 What did we learn?

1. Why are algae such important organisms?

2. What gives algae its green color?

 **Taking it further**

1. Why are some algae yellow, brown, blue, or red?

| God's Design: Life | The World of Plants | Day 56 | Unit 6 Lesson 34 | Name

34 Fungi

Are these really plants?

🧪 Supply list

☐ 6 slices of bread (homemade works best)

☐ 3 plastic sandwich bags or plastic zipper bags

☐ Copy of "Mold Data Sheet"

🧪 Grow That Mold

Follow the instructions on page 130 of the student book. Be sure to write your observations over a number of days on the Mold Data Sheet found on the following page. What did you learn from the activity?

🏅 Supplies for Challenge

☐ Fresh mushroom

☐ Index card

☐ Aerosol hairspray

🏅 Spore Print Challenge

Follow the instructions on page 131 of the student book. **Note**: It is important to leave the cap on the index card overnight. Were you able to make and preserve the spore print? What was the most interesting thing you observed?

🧠 What did we learn?

1. Why are fungi not considered plants and given their own kingdom?

2. What are some good uses for fungi?

🚀 Taking it further

1. What other conditions might affect mold growth other than those tested here?

2. How can you keep your bread from becoming moldy?

Name _____   Date _____

🧪 Mold Data Sheet

Hypothesis: I think that mold will grow best on the slice(s) that is (are) _____
_____

Date	Warm/dry No bag	Warm/dry In a bag	Cold/dry No bag	Cold/dry In a bag	Warm/moist No bag	Warm/moist In a bag

Did your data match your hypothesis? _____

What conditions were best for mold growth? _____

| | God's Design: Life | The World of Plants | Day 59 | Unit 6 Lesson 35 | Name |

 **Conclusion**

Appreciating the world of plants

 **Supply list**

☐ Tagboard

☐ Dried leaves, flowers, grass

☐ Glue

☐ Seeds, seed pods, other parts of plants

Make a Plant Collage

Using these materials and glue, make a unique and beautiful picture that reminds you of something from God's wonderful world of plants. You can also write a poem that expresses your wonder at the amazing design of plants.

Anatomy Worksheets

for Use with

The Human Body
(God's Design: Life Series)

God's Design: Life | The Human Body | Day 61 | Unit 1 Lesson 1 | Name

The Creation of Life

God created them male and female.

Supply list

☐ Paper

☐ Colored pencils

☐ Mirror

☐ Bible

Self-portrait Activity

Follow the instructions on page 141 of your student book, reading and discussing the Scripture and discussing what the Bible says about why we were created and how much God loves us. Make sure that you do the self-portrait of yourself.

Challenge: Body Systems

This challenge is to determine what you already know about the systems of the human body. Learn more in the instructions for this challenge on page 141.

1. Which system do you know the most about?

2. Which system do you know the least about?

3. Which system is the most interesting to you?

What did we learn?

1. On which day of creation did God make man?

2. In whose image did God create man?

3. According to Genesis 1:26, over what were man and woman to rule?

 **Taking it further**

1. Since we are created in God's image, how should we treat our bodies?

| God's Design: Life | The Human Body | Day 62 | Unit 1 Lesson 2 | Name |

2 Overview of the Human Body

We are fearfully and wonderfully made!

Supply list

☐ Copy of "Body Wheel"
☐ Crayons, colored pencils, or markers
☐ Paper fasteners
☐ Scissors

Body Wheel Activity

You will be assembling a "Body Wheel" using the instructions found on page 143. The Body Wheel parts can be found on pages 119 and 121 of this teacher guide.

Other Systems Challenge

Review your list from lesson 1 and then add any you missed to it. Learn more about three other systems of the body on page 143 of the student book. Add these systems to your list and include a brief description of each. How does God's design of our bodies impact how they work? Discuss your answer to this question with your teacher.

What did we learn?

1. Name as many of the body's systems as you can and describe what each system does.

The Human Body // 117

🚀 Taking it further

1. Which body systems are used when you walk across a room?

Name _____ Date_____

🧪 Body Wheel
Part 1

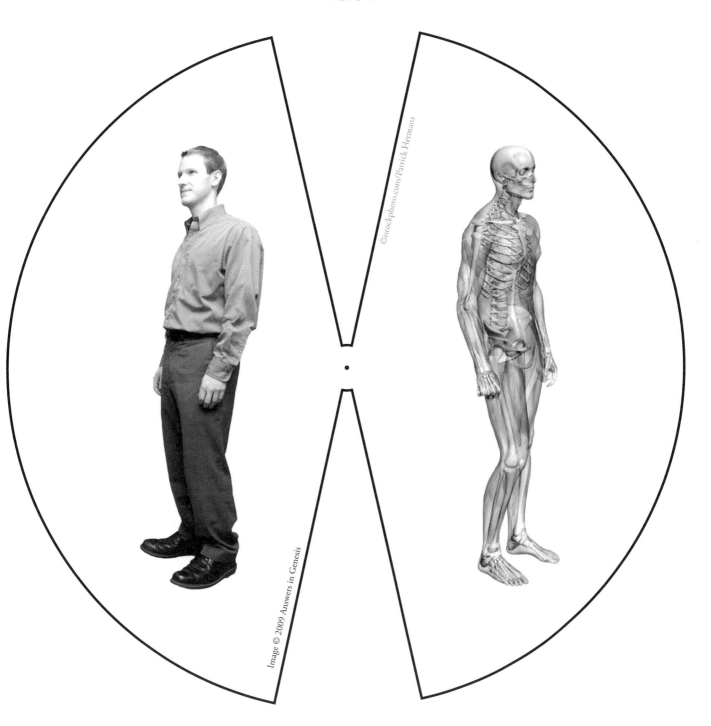

The Human Body // 119

Body Wheel
Part 2

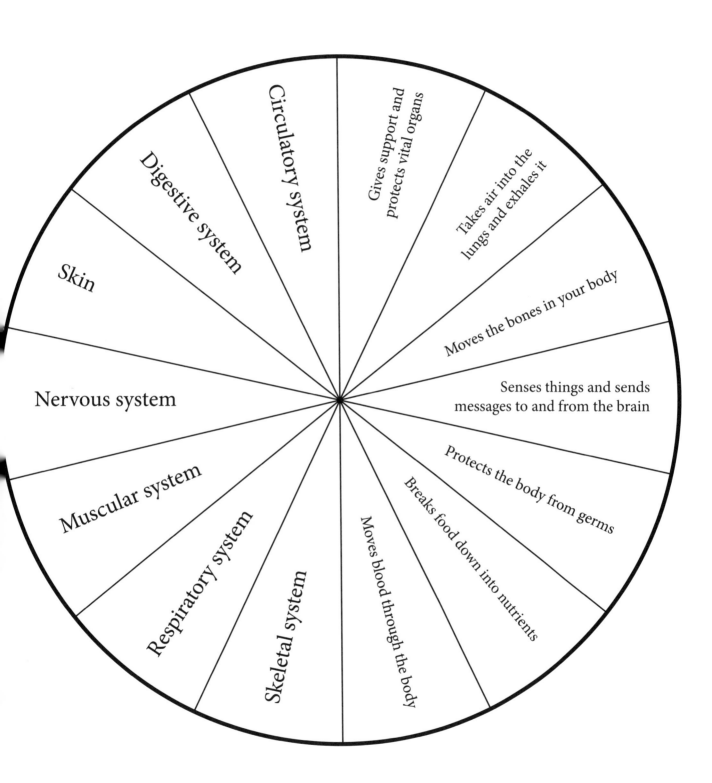

The Human Body 121

| God's Design: Life | The Human Body | Day 64 | Unit 1 Lesson 3 | Name |

3 Cells, Tissues, & Organs

The building blocks of our bodies

Supply list

☐ Copy of "Body Cells" worksheet

Body Cells Worksheet

Complete the Body Cells Worksheet on page 125 of this teacher guide. You can use the image on page 147 of the student book as a guide.

Challenge: Tissue Types

If you have read page 148 in the student book, then you have just learned about four different categories of tissues. On the list below, identify which type of tissue each body part belongs to.

Skin:

Muscles:

Tendons:

Lining of the mouth:

Brain:

Inside of lungs:

Fat:

Bones:

What did we learn?

1. What is the function of each of the following kinds of cells: skin cells, red blood cells, white blood cells, bone cells, nerve cells, and muscle cells?

Taking it further

1. How has God uniquely designed red blood cells to transport oxygen?

2. How are nerve cells specially designed to carry signals?

3. How did God design skin cells to perform their special functions?

4. With all these cells working together, what do you think is the largest organ in the body?

Name _____   Date _____

🧪 Body Cells Worksheet

Write the type of cell on the line next to the description of that cell. Choose the names from the list of cells below.

Nerve cell	White blood cell	Skin cell
Red blood cell	Bone cell	Muscle cell

A. _____ I am designed to carry messages quickly throughout the body.

B. _____ I stretch and contract to allow for movement.

C. _____ I attack invading germs and surround them so they cannot make the body sick.

D. _____ I help keep moisture inside and germs outside the body.

E. _____ I carry oxygen to all the parts of the body and carry carbon dioxide away.

F. _____ I give the body strength and form.

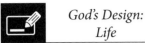

God's Design: Life | The Human Body | Day 67 | Unit 2 Lesson 4 | Name

4 The Skeletal System
Structure and strength

Supply list
☐ Sandy Skeleton (If you are making a copy of Sandy Skeleton, do not copy back to back.)
☐ Scissors
☐ 5 paper fasteners for each child

Sandy Skeleton

Cut out and assemble Sandy Skeleton found on pages 129 and 131 of this teacher guide. Sandy Skeleton can be put together without the cutouts being done perfectly by the student. For younger students, the instructor can cut out the pieces for the student to assemble.

Supplies for Challenge
☐ Tape measure

Your Skeleton Challenge

Read more details about the skeletal system on page 152 of your student book and then take this challenge. Carefully measure your height first thing in the morning. Then, measure your height again just before you go to bed and see if you notice a difference.

Morning Height: _____

Evening Height: _____

Keep your assembled Sandy Skeleton — you will be using it in an upcoming lesson.

What did we learn?

1. What are three jobs that bones perform?
   a.
   b.
   c.

2. How are muscles connected to bones?

3. What keeps bones from rubbing against each other at the joints?

4. How many bones does an adult human have?

5. What is the main mineral in bones?

Taking it further

1. What do you think is the largest bone in the body?

2. Why does this bone need to be so large?

3. What do you think are the smallest bones in the body?

Name _____   Date _____

🧪 Sandy Skeleton

Keep assembled skeleton for use in future lessons.

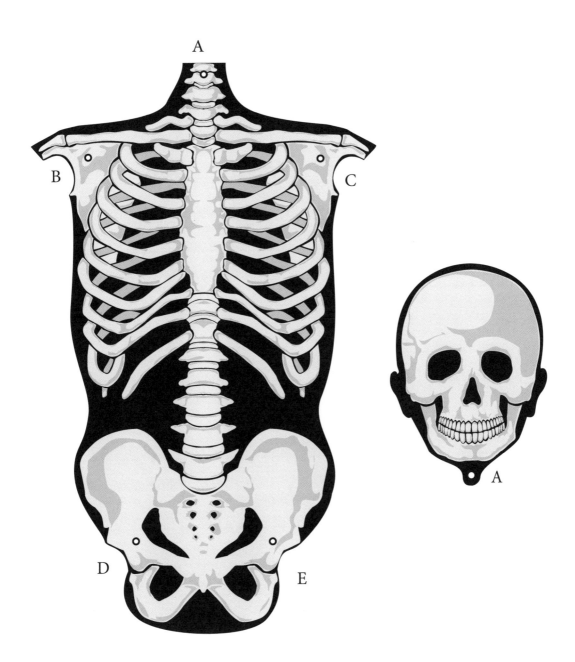

The Human Body 129

🧪 Sandy Skeleton

Keep assembled skeleton for use in future lessons.

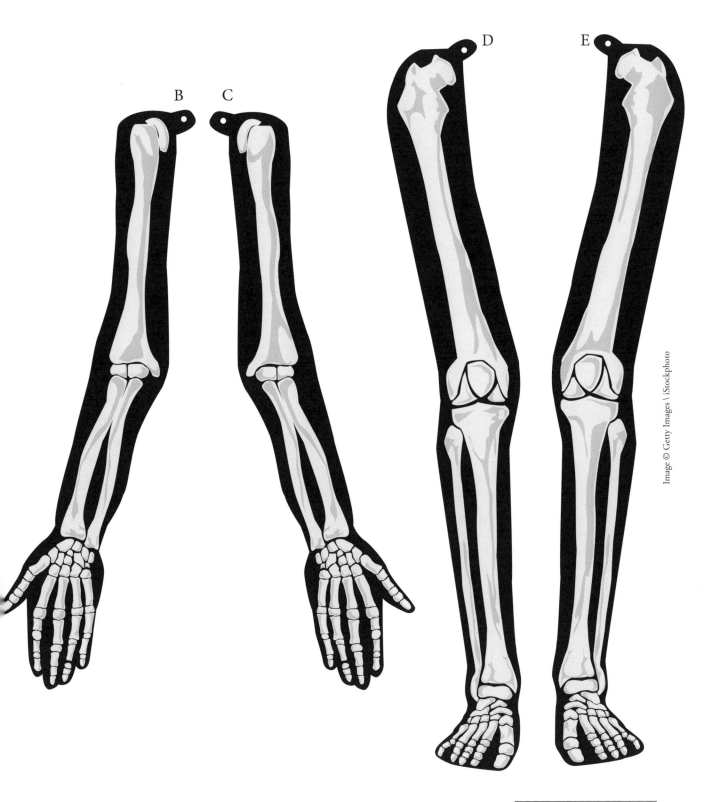

The Human Body // 131

| God's Design: Life | The Human Body | Day 68 | Unit 2 Lesson 5 | Name |

5 Names of Bones

What's a clavicle?

🧪 Supply list

☐ Washable gel pens or sticky notes
☐ Anatomy book

🧪 Label Those Bones Activity

Directions for this activity are found on page 155 of the student book. If you choose not to use the erasable pen, the sticky notes will also work for the activity. Be sure to make up a song about bones when you finish labeling them.

🏅 Supplies for Challenge

☐ Anatomy book
☐ Copy of "What's My Name?" worksheet

🏅 What's My Name?

Complete this challenge using the What's My Name? Worksheet on page 135 of this teacher guide.

🧠 What did we learn?

1. Review the names of the bones by pointing to each bone as you name it.

2. Is your cranium above or below your mandible?

The Human Body // 133

3. What is moving if you wiggle your phalanges?

🚀 Taking it further

1. What happens if you cross your legs and gently hit just below your patella?

2. Why do we have Latin names for body parts?

Name _____ Date _____

🎖 What's My Name? Worksheet

Try to solve these riddles with the name of a bone in your body from the list below. Use an anatomy book to help you.

Patella	Clavicle	Femur	Radius
Sternum	Scapula or vertebrae	Hammer, anvil, stirrup	
Phalanges	Mandible	Fibula	

1. I hold your ribs together. _____

2. Your collar rests on me. _____

3. Pat me on the back. _____

4. I support your weight when you stand. _____

5. Listen, you might find us in a blacksmith shop. _____

6. I'm in your leg, and that's no lie. _____

7. I rotate around your wrist. _____

8. You wouldn't want to stub me. _____

9. Man, I talk a lot. _____

10. When you cross your legs, I pop up. _____

The Human Body

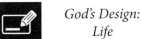

| God's Design: Life | The Human Body | Day 69 | Unit 2 Lesson 6 | Name |

6 Types of Bones

Are all bones created equal?

 Supply list

☐ Anatomy book
☐ Model of Sandy Skeleton from lesson 4

 Simon Says

This is a great way to review the names of the bones you have learned. Follow the directions on page 157 of your student book, and have fun learning while you play Simon Says.

Supplies for Challenge

☐ 2 chicken leg or thigh bones
☐ Vinegar
☐ Cup

Broken Bones Challenge

Read the text on page 158 of your student book to learn about how the body repairs itself when we break a bone. Then follow the instructions on the challenge to learn what gives bones their strength. Be sure to share with your teacher what you learned from doing the challenge.

What did we learn?

1. Which bones are designed mainly for protection of internal organs?

2. Which type of bones help determine what your face will look like?

3. Which type of bones work closely with your circulatory system to replace old blood cells?

🚀 Taking it further

1. Why are the long bones filled with marrow and not solid?

2. What is the advantage of having so many small bones in your hands?

God's Design: Life | The Human Body | Day 71 | Unit 2 Lesson 7 | Name

 Joints

Connections are important.

🧪 Supply list

☐ Items found around your house
- ☐ sliding doors
- ☐ pets
- ☐ LEGO® pieces
- ☐ nut crackers
- ☐ pliers

🧪 Scavenger Hunt

On page 160 of your student book, you will find a list of different joints to find around your house. See how many you can find and note any not on the list that you also found.

🏅 Supplies for Challenge

☐ 2 wooden pencils
☐ Tacks
☐ Wide rubber bands

🏅 Amazing Joints

This challenge is to build a model of a joint. The procedure is on page 161 of your student book. Show your teacher the finished joint. Take a picture and share with others what you have learned!

🧠 What did we learn?

1. What was the most common joint found around your house?

🚀 Taking it further

1. Which came first, the joints in the body or the joints in your house?

2. Why do you need so many different kinds of joints in your body?

God's Design: Life | The Human Body | Day 72 | Unit 2 Lesson 8 | Name

8 The Muscular System

Making it move

 Find the Pairs

It's time to see how many muscle pairs you can find on your body. Using the image and instructions on page 164 of the student book as an example, see how many muscle pairs you can find on yourself. How many did you find?

Supplies for Challenge

☐ Piece of raw steak or other meat
☐ Magnifying glass

Muscle Jobs

Read about the jobs that muscles have in your body on page 164 of the student book. Use a magnifying glass to closely examine a piece of raw steak. Look for parallel lines of meat. These are the muscle tissues that are lined up parallel to allow the muscles to contract in a strong way. On a piece of paper, draw an image of the muscles that you see on the raw steak.

What did we learn?

1. How does a contracted muscle feel?

2. How does a muscle get stretched?

🚀 Taking it further

1. How does a muscle know when to contract?

2. How does your face express emotion?

| God's Design: Life | The Human Body | Day 73 | Unit 2 Lesson 9 | Name |

9 Different Types of Muscles

Aren't they all the same?

Supply list

☐ Yard stick
☐ Stopwatch

Muscle Memory Exercises

Practice the activities on page 166 of your student textbook every day for three days. You will notice that your muscles "gain a memory" and improve from the first time to the last time.

Challenge: Muscle Tissue

Think about the purpose of each of the muscles listed below. Decide if they are likely to be made of striated, cardiac, or smooth muscle tissue. Write your answer by each muscle.

1. Diaphragm:

2. Tongue:

3. Esophagus:

4. Mother's womb:

5. Hand muscle:

6. Heart:

What did we learn?

1. What are the two types of muscles?

2. How can we keep our muscles healthy?

3. How do your muscles learn?

4. What are some advantages of exercising?

Taking it further

1. Do you need to exercise your facial muscles?

| God's Design: Life | The Human Body | Day 74 | Unit 2 Lesson 10 | Name |

10 Hands & Feet

Special designs from God

Supply list

☐ Pencil
☐ Paper
☐ Cup

Design Appreciation

It's kind of amazing just how well-designed your hands and feet are! Learn more with the fun activities on page 168 of your student book. Tell your teacher how each one worked out!

Supplies for Challenge

☐ Anatomy book
☐ Drawing materials

Challenge: Right- or Left-handed?

Which hand do you use to write with and quickly use to pick up things? Read about how the brain is involved with which hand you use to do daily tasks. Do a short presentation about right- and left-handedness.

What did we learn?

1. Which is the most important finger?

2. Why is the thumb so important?

3. What are some special features God gave to hands and feet?

🚀 Taking it further

1. What activities or jobs require special use of the hands?

2. What jobs require special use of the feet?

| | God's Design: Life | The Human Body | Day 77 | Unit 3 Lesson 11 | Name |

11 The Nervous System
Telling your body what to do

🧪 Supply list

☐ Copy of "Nervous System Coloring Page"

☐ Stopwatch

☐ Ruler

☐ Small object such as an eraser, pebble, or small toy

🧪 Response Time Test

Follow the procedure on page 172 of the student book to determine how quickly a message can go from the eyes to the brain and then to the hand to cause it to move out of the way.

🧪 Nervous System Coloring Page

Complete the Nervous System Coloring Page on page 149 of the teacher guide. Follow the instructions on the coloring page carefully and label it as needed.

🏅 Challenge: Unique Humans

Make your own list of things that humans can do that animals cannot do. What accounts for each of these abilities?

What did we learn?

1. What are the three main parts of the nervous system?

    a.

    b.

    c.

2. In the response time test, what messages were sent to and from the brain?

Taking it further

1. Name ways that information is collected by your body to be sent to the brain.

Name _____   Date _____

🧪 Nervous System Coloring Page

Color the brain blue, the spinal cord green, and the nerves red. Label the central and peripheral nervous systems.

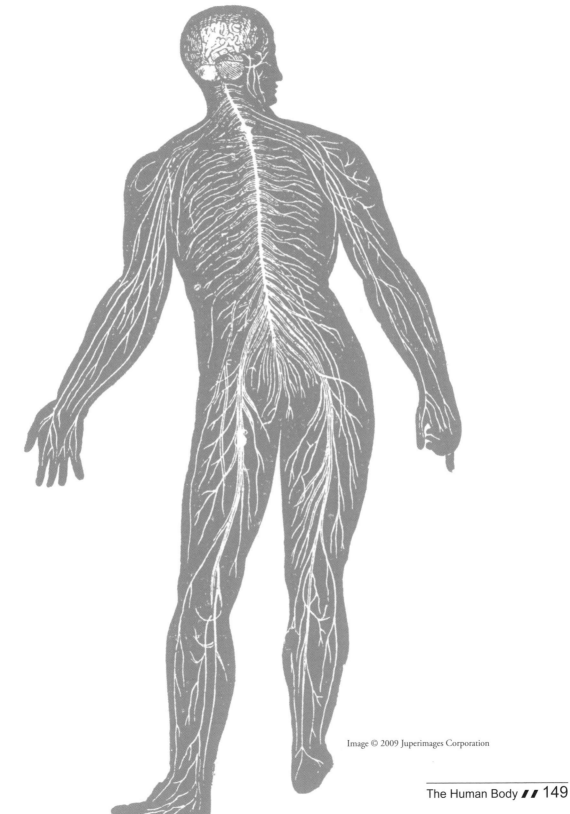

Image © 2009 Juperimages Corporation

The Human Body // 149

| God's Design: Life | The Human Body | Day 78 | Unit 3 Lesson 12 | Name |

12 The Brain

Captain of the ship

🧪 Supply list

☐ 3 different colors of modeling clay

☐ Anatomy book

🧪 Brain Model

Use modeling clay to make a model of the brain. You can use the picture on page 175 (top left) of your student book as a guide.

🏅 Brain Anatomy Challenge

You will find two lists and instructions on page 176 of your student book. The first is a list of additional parts of the brain that you can add to your model. Be sure to label your model with all the different parts. Use the second list to figure out in which part of the brain each function listed takes place.

1. Thought:

2. Smell:

3. Heart beat regulation:

4. Memory:

5. Sight:

6. Speech:

7. Muscle control:

8. Pupil dilation:

What did we learn?

1. What are the three major parts of the brain?

    a.

    b.

    c.

2. Which part of the brain controls growth?

Taking it further

1. Which part of the brain would be used for each of the following?

    - running

    - dilating your eyes

    - learning your math facts

2. Is your brain the same thing as your mind?

| *God's Design: Life* | The Human Body | Day 79 | Unit 3 Lesson 13 | Name |

13 Learning & Thinking

How do you use your brain?

 Supply list

☐ 6 index cards

☐ 6 different colored markers

 Brain Exercises

Follow the activity instructions on page 178 of your student book. What did you learn in this activity? Explain it to your teacher.

 Challenge: Logic Puzzles

1. Four people must cross a river in a boat. Two people weigh 50 pounds each and the other two weigh 100 pounds each. They have a boat that can only hold a maximum of 100 pounds without sinking. Describe how all four people can cross the river in the boat.

2. You are in a strange land and you need to find the nearest town. You know that one group of people living in the area always tells the truth and that the second group of people always tells a lie. You come to a fork in the road and do not know which way to go. Standing at the fork are two people, one from each group, but you cannot tell which one is from which group. You can only ask one person one question. What question will you ask to be sure you take the correct road to reach the nearest town?

3. Create your own logic puzzle.

What did we learn?

1. Which part of the brain does each of the following:

    - stores short-term memories

    - stores long-term memories

    - controls learning and thinking

    - controls the senses?

2. Which side of the brain controls the left side of the body?

3. What is necessary for a healthy brain?

Taking it further

1. List ways you can learn something.

2. What is something you have trouble learning?

| God's Design: Life | The Human Body | Day 82 | Unit 3 Lesson 14 | Name |

Reflexes & Nerves

Faster than lightning

Supply list

☐ 2 sharp pencils or 2 toothpicks
☐ Blindfold

Test Your Reflexes & Nerves

Follow the instructions on page 183 of your student book to test your own reflexes. **Note:** Do this with your teacher or with another student with adult supervision. Be sure the participants are instructed to be gentle when using the toothpicks or pencils to do the test.

Supplies for Challenge

☐ Drawing materials
☐ Research materials on multiple sclerosis

Challenge: Nerve Cells

Read the text on page 184 of the student book. Then on a piece of paper, draw a diagram of a nerve cell, labeling all the parts. Bonus! Do some research on multiple sclerosis, a disease of the nervous system. Find out what causes it and how it affects the nerves. Write a short report or do a short oral presentation of what you learned.

What did we learn?

1. How do reflex reactions differ from other nervous system messages?

2. Why do we have reflexes?

3. What are some different types of sensations detected by your nerves?

🚀 Taking it further

1. What reflexes might you experience?

2. How does the sense of touch differ from your fingertips to the back of your arm?

3. Why do you need a larger number of nerves on the bottoms of your feet?

| 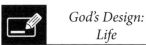 God's Design: Life | The Human Body | Day 83 | Unit 3 Lesson 15 | Name |

15 The Five Senses

Letting your brain know what's out there

Supply list

- ☐ 3 bowls
- ☐ Hot water
- ☐ Warm water
- ☐ Cold water
- ☐ Sandwich bag filled with ice
- ☐ Jacket with a zipper
- ☐ 2–3 straight pins

Feeling Hot, Feeling Cold

Are you ready to learn how our bodies react to temperature differences? Do the activity procedures on page 186 of your student book. What did you learn? **Note**: When using hot water, be sure it is not too hot to the touch. Ask for assistance from an adult if needed.

Supplies for Challenge

- ☐ Glue
- ☐ Paper

Challenge: Braille System

Why do you think that Braille was designed to be read with fingertips instead of your palm or the side of your hand? What message did you make for the challenge?

What did we learn?

1. What are your five senses?
   a.
   b.
   c.
   d.
   e.

2. Which of these senses usually gives us the most information?

3. How does your brain compensate for the loss of one of your senses?

🚀 Taking it further

1. You have nerves all over your skin, so why don't you feel your clothes all day long?

2. Your eyes see your nose all day long. Why don't you notice it all the time?

3. If you are in the hot sun for a while then you go inside, the room feels cold. Why?

| God's Design: Life | The Human Body | Day 84 | Unit 3 Lesson 16 | Name |

16 The Eye

Window to the world

🧪 Supply list

☐ Paper tube (such as a paper towel roll or a rolled piece of paper)

☐ Piece of paper

🧪 Fooling Your Eyes

Do the activity on page 190 of your student book to help understand how your eyes work. There are four activities that can be done. You can do all four or only a few depending on the requirements of your teacher.

🏅 Supplies for Challenge

☐ Anatomy book

🏅 Challenge: Liquid in Your Eyes

Read the information for this challenge on page 190 of the student book and you can use the diagram on page 189 as a reference. On a sheet of paper, draw a detailed diagram of the eye, labeling the following parts:

- Lens
- Pupil
- Iris
- Cornea
- Rods
- Cones
- Retina
- Optic nerve
- Vitreous humor
- Aqueous humor

What did we learn?

1. Name four important parts of the eye.
   a.
   b.
   c.
   d.

2. How does your brain compensate for different amounts of light in your surroundings?

3. How does your brain help you to focus on items that are near and items that are far away?

4. Why did God design our bodies with two eyes instead of just one?

5. How does having two eyes help with a 3-dimensional image?

6. Since you have a blind spot, how can you see what is in that spot?

Taking it further

1. Name some ways that the eye is protected from harm.

2. Why do some people have to wear glasses or contact lenses?

3. Why can you fool your eyes or your brain into thinking you saw something you didn't actually see?

| God's Design: Life | The Human Body | Day 86 | Unit 3 Lesson 17 | Name |

17 The Ear

Do you hear what I hear?

🧪 Supply list

☐ Copy of "Do You Hear What I Hear?" worksheet

🧪 Do You Hear What I Hear?

Complete the Do You Hear What I Hear? Worksheet on page 162.

🏅 Ear Anatomy Challenge

Read the text on page 193 of your student book. Describe to your teacher what you learned about how the ear functions in relation to balance and movement.

🧠 What did we learn?

1. What characteristic of a sound wave determines how high or low a sound will be?

2. What characteristic of a sound wave determines how loud or soft a sound will be?

🚀 Taking it further

1. Why do two different instruments playing the same note at the same loudness sound different?

2. Name several ways to protect your hearing.

3. How do you suppose deaf children learn to speak?

4. How do you think a CD player or a telephone makes sounds?

🧪 Do You Hear What I Hear? Worksheet

The higher the pitch or frequency is, the closer together the sound waves will be. Write the letter of the sound wave that demonstrates the most likely frequency next to each item listed.

____ Steaming tea kettle   A.

____ Violin   B.

____ Man singing   C.

____ Bass drum   D.

The louder a sound is, the taller or bigger the sound wave's amplitude will be. Write the letter of the sound wave that demonstrates the most likely amplitude next to each item listed.

____ Falling snow   A.

____ A jet engine   B.

____ A TV show   C.

____ A whisper   D.

God's Design: Life | The Human Body | Day 87 | Unit 3 Lesson 18 | Name

18 Taste & Smell

What's for dinner?

🧪 Supply list

☐ Paper towels
☐ Salt
☐ Sugar
☐ Potato
☐ Apple
☐ Carrot
☐ Several spices or other items with familiar smells (cinnamon, peppermint, lemon juice, vinegar)

🧪 Taste Without Saliva?

Follow the instructions on page 195 of the student book. What did you discover? If you have time, complete the following activity.

🧪 Taste Without Smell

Follow the instructions on page 195. If doing the activity with younger students, the instructor can peel and cube the fruit and the vegetables. Be careful using the knife.

How difficult was it to tell the difference between the foods without using your nose? Unplug your nose. Was it easier for you to tell which food was which?

Bonus Activity! If you have time, you can also do the Smell Detective activity. See how many things you can identify just by smell.

🏅 Challenge: How We Taste and Smell

Read the text on page 196 of your student book. Tell your teacher what you learned about how your senses help you understand the world around you.

🧠 What did we learn?

1. What four flavors can your tongue detect?

   a.

   b.

   c.

   d.

2. How does your tongue detect flavors?

3. How does your nose detect fragrances?

🚀 Taking it further

1. Can you still taste foods when you have a stuffy nose?

2. Smells are used for more things than just enjoying food. List some other uses for your sense of smell.

3. Oranges and grapefruits are both sweet and sour. Why do they taste different?

4. Cocoa is very bitter. Why does chocolate candy taste so delicious?

| God's Design: Life | The Human Body | Day 89 | Unit 4 Lesson 19 | Name |

19 The Digestive System

What happens to my lunch?

🧪 Supply list

☐ Sandwich
☐ Clock
☐ Copy of "Where's My Lunch?" worksheet

🧪 Where's My Lunch?

Complete the Where's My Lunch? Worksheet on page 167 of this teacher guide. Be sure to carefully read the instructions for this activity on page 199 of the student book. It may be interesting to follow the path of your lunch using the diagram also found on the page.

🏅 Supplies for Challenge

☐ Anatomy book
☐ Copy of "Digestive System" worksheet

🏅 Chemicals Challenge

Using an anatomy book or the diagram on page 199 of the student book, label all of the parts of the digestive system on the "Digestive System" worksheet. It is found on page 168 of the teacher guide.

🧠 What did we learn?

1. What are the main parts of the digestive system?

2. What role do your teeth play in digestion?

3. What role does your tongue play in digestion?

4. Which is longer, your small intestine or your large intestine?

5. Which is wider, your small intestine or your large intestine?

🚀 Taking it further

1. Can you eat or drink while standing on your head?

2. Why do some foods spend 30 minutes in the stomach while other foods spend 3 hours in the stomach?

3. What makes you feel hungry?

4. Why did God design your body with a way to make you feel hungry?

Name _____   Date _____

🧪 Where's My Lunch? Worksheet

Because you cannot tell when your food actually enters or exits a particular part of the digestive system, you have to estimate, or make a good guess. This exercise will help make you more aware of what your food is doing for you.

Location	Estimated time entered	Estimated time exited
Mouth Enter the time you took the first bite as the time entered and the time you swallowed the last bite of your lunch as the time exited.		
Esophagus Food spends about 10 seconds in your esophagus.		
Stomach Food can spend 2–3 hours here. If you ate peanut butter and jelly or lunch meat, estimate 3 hours.		
Small intestine It takes food about 8 hours to pass through this long tube.		
Large intestine It takes 24–48 hours for food to pass through this 4–5-foot long tube.		

Name _____   Date _____

🎖 Digestive System Worksheet

Label the following parts of the digestive system with the correct letters from the drawing.

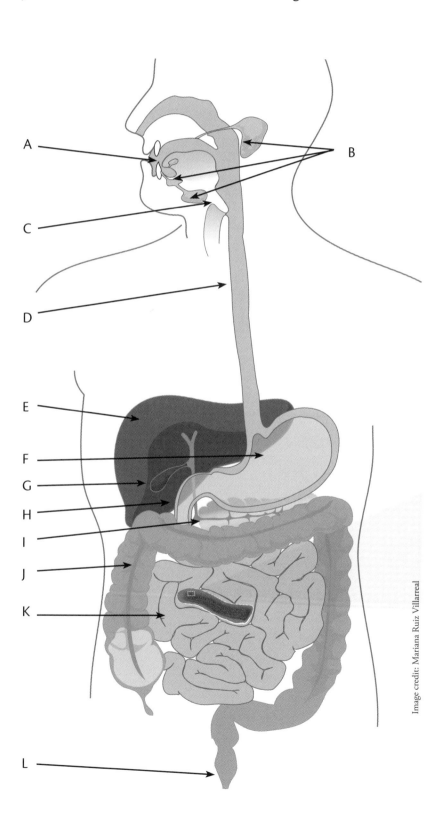

1. _____ Pancreas
2. _____ Salivary glands
3. _____ Tongue
4. _____ Anus
5. _____ Esophagus
6. _____ Small intestine
7. _____ Stomach
8. _____ Liver
9. _____ Gall bladder
10. _____ Duodenum
11. _____ Epiglottis
12. _____ Large intestine

| God's Design: Life | The Human Body | Day 91 | Unit 4 Lesson 20 | Name |

Teeth

Grind that food.

🧪 Supply list

- ☐ Tagboard, poster board, or other thick paper
- ☐ Modeling clay
- ☐ Plaster of Paris
- ☐ Foil
- ☐ Tape
- ☐ Scissors
- ☐ Small bowl or cup
- ☐ Spoon

🧪 Teeth Molds Activity

Note: This activity requires adult supervision. Follow the procedure found on page 202 of the student book. When completed, be sure to name each tooth and review its function in the eating process.

🎖 Supplies for Challenge

- ☐ Three colors of modeling clay

🎖 Tooth Structure Challenge

Use three different colors of modeling clay to make a cross-section model of a tooth. Be sure to review the function and name of each part of the tooth.

🧠 What did we learn?

1. What is the job of each kind of tooth?

2. Why do people have baby teeth and why do they fall out?

🚀 Taking it further

1. Why do we need to take care of our teeth?

God's Design: Life | The Human Body | Day 92 | Unit 4 Lesson 21 | Name

21 Dental Health

Taking care of those teeth

Supply list

☐ Toothbrush
☐ Toothpaste
☐ Dental floss
☐ Mirror

Practice, Practice

Review how you properly brush your teeth by completing the activity on page 204 of the student book.

Straight Teeth

Do a survey to find out how many people you know have had braces. Ask 10 adults and 10 young people and find out how many of them have had or currently have braces. What were your findings? Which group had more people that had or currently have braces?

What did we learn?

1. What are three things you can do to have healthy teeth?

2. How does brushing your teeth help keep them healthy?

3. List some foods that are good for your teeth.

4. List some foods that are bad for your teeth.

 Taking it further

1. Since your baby teeth are going to fall out anyway, why do you still need to brush them and take care of them?

| God's Design: Life | The Human Body | Day 93 | Unit 4 Lesson 22 | Name |

22 Nutrition

What should you eat?

Supply list

☐ Foods from all of the food groups
 (including breads/grains, fruits, vegetables, dairy, and meat/nuts/dried beans)

Eating a Healthy Meal

Read the instructions on page 208 of your student book. Now plan a healthy meal for your family. Be sure to select foods from all the different food groups.

Supplies for Challenge

☐ Copy of "Nutrition" worksheet

Challenge: Nutrition Worksheet

Complete the Nutrition Worksheet on page 174 of this teacher guide. This will help you learn more about the types of food that are good and bad for you. Hint! Helpful information can often be found on food labels.

What did we learn?

1. What are the five food groups listed in this lesson?

2. What types of foods should you eat only a small amount of each day?

3. Why is variety in your diet important?

Taking it further

1. Can a vegetarian eat a balanced diet? Hint: What other foods contain proteins found in meat?

2. Is it necessary to eat dessert to have a healthy diet?

Name _____ Date_____

🧪 Nutrition Worksheet

Complete as much of the chart below as you can, then answer the questions at the bottom of the page.

	Banana	Cup of yogurt	Fast food hamburger	French fries	Candy bar	Spaghetti & tomato sauce
Serving size						
Calories						
Grams of fat						
Grams of carbohydrates						
Grams of protein						
Milligrams of sodium (salt)						
Major vitamins and minerals						
Grams of cholesterol						
Amount of fiber						
Other important nutrients						

Which food has the most fat in it? _____

Which food has the most calories? _____

Which food has the most salt? _____

Which food do you think would be healthiest for you? _____

| God's Design: Life | The Human Body | Day 96 | Unit 4 Lesson 23 | Name |

23 Vitamins & Minerals

Do I have to go to a mine to get minerals?

Supply list

☐ Food in your kitchen

☐ Paper

☐ Pencil

Vitamin & Mineral Scavenger Hunt

Find as many of the vitamins and minerals listed in this lesson as you can. Follow the instructions on page 212 of the student book. Pages 211 and 212 contain the list of vitamins and minerals you are looking for.

Health Problems Challenge

Read the text on page 213 of the student book. Choose one of the health problems discussed as the focus of a presentation on the need to eat healthy, or you can do a poster with information warning people about the dangers of not eating foods with important vitamins and minerals. Be creative in the way you choose to share what you know with your teacher or others.

What did we learn?

1. What are the three main forms of energy found in food?

    a.

    b.

    c.

2. How can we be sure to get enough vitamins and minerals in our diet?

The Human Body 175

3. Why is water so important to our diet?

🚀 Taking it further

1. Can you drink soda instead of water?

2. Are frozen dinners just as healthy as fresh food?

3. Is restaurant food as healthy as home-cooked food?

| God's Design: Life | The Human Body | Day 98 | Unit 5 Lesson 24 | Name |

24 The Circulatory System
The transportation highway

Supply list
☐ Stopwatch

Measuring Your Pulse

Follow the instructions on page 217 of your student book to test the effects of exercise on your heartbeat. What did you learn?

Blood Pressure

If you have a chance, find out what your blood pressure is. You can have your blood pressure measured at a doctor's office or clinic. Some people have their own blood pressure machines. If you know someone with one, ask them to take your blood pressure. If you cannot find a way to get your blood pressure taken, do a presentation that explains how the process of taking blood pressure works, or you could do a diagram that explains it.

What did we learn?

1. What are the three main parts of the circulatory system?

    a.

    b.

    c.

2. What are two functions of blood?

    a.

    b.

3. What are three types of blood vessels?

    a.

    b.

    c.

4. Which blood vessels carry blood away from the heart?

5. Which blood vessels carry blood toward the heart?

6. What happens to the blood in the capillaries?

🚀 Taking it further

1. How is the circulatory system like a highway?

2. Why is exercise important for your circulatory system?

3. List two other systems that depend on the circulatory system to function properly.

4. Why does your pulse increase when you exercise?

| God's Design: Life | The Human Body | Day 99 | Unit 5 Lesson 25 | Name |

25 The Heart
Master pump

Supply list

☐ Copy of "The Heart" worksheet

☐ Blue and red colored pencils

Blood Flow in the Heart

Complete the Heart Worksheet on page 181 of the teacher guide. Be sure to use the correct colors to show the difference in blood with oxygen or with carbon dioxide.

Supplies for Challenge

☐ Cow's heart (beef heart) or sheep's heart (may be fresh or preserved)

☐ Rubber gloves

☐ Sharp knife or scalpel

☐ Anatomy book or dissection guide

Examine a Heart

Note: This is an optional challenge. Science kits are available online from various companies for this type of activity if you wish to complete it. Use the knife very carefully for this activity if you do it.

If you do not want to do the actual dissection, there are videos available on YouTube of cow or sheep hearts being dissected. It is very important that the teacher prescreen any video to make sure it is age appropriate.

What did we learn?

1. What are the four chambers of the heart?

    a.

    b.

    c.

    d.

2. How many times does a blood cell pass through the heart on each trip around the body?

🚀 Taking it further

1. What are some things you can do to help your heart stay healthy?

2. Is your heart shaped like a valentine?

3. Does Jesus live in your physical heart?

Name _____ Date _____

🧪 The Heart Worksheet

Label where the blood is flowing from or to on each line. Then color each part to indicate if it contains blood with oxygen or blood without oxygen.

Red = blood with oxygen

Blue = blood with carbon dioxide (no oxygen)

A. _____

B. _____
Superior Vena Cava

C. _____
Pulmonary Artery

D. _____
Pulmonary Vein

E. _____ F. _____

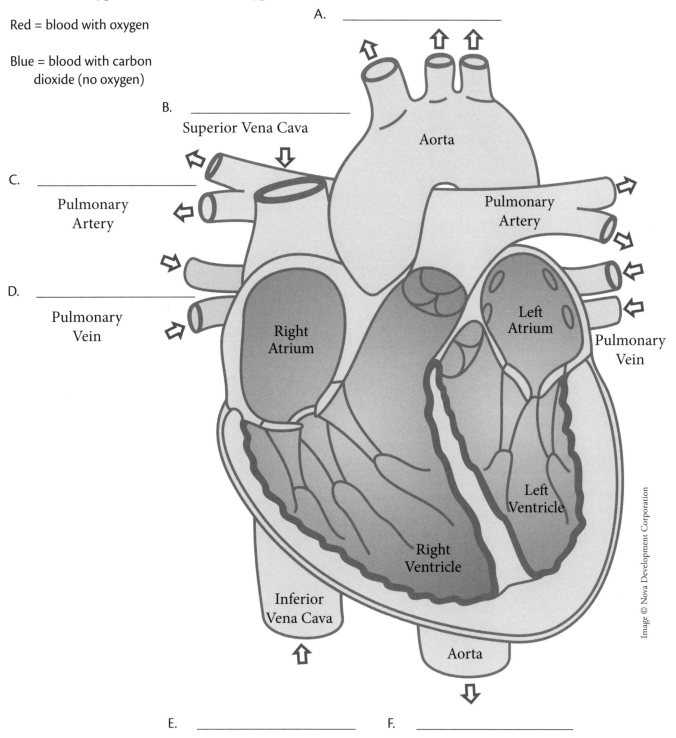

The Human Body // 181

| | God's Design: Life | The Human Body | Day 101 | Unit 5 Lesson 26 | Name |

Blood

Delivery trucks and policemen

🧪 Supply list

☐ Several chairs
☐ Several children (if available)
☐ Red Hots candies
☐ Corn syrup

☐ White jelly beans
☐ Candy sprinkles
☐ Red and blue construction paper

🧪 Blood Cell Game

This is a great way to model the role of blood cells in the body. Follow the instructions on page 223 of the student book.

🧪 Making "Sample" Blood

Use candies to make a representation of a blood sample. What is the main color of your mixture?
Note: This may get messy so be sure to use a big enough bowl!

🏅 Supplies for Challenge

☐ Copy of "Blood Types" worksheets 1 & 2

🏅 Blood Transfusions

Read the information on page 224. Then fill out the Blood Types 1 Worksheet on page 185 of the teacher guide, showing which type or types of blood can donate to each blood type and which type or types of blood can be received by each blood type.

Next, fill out the Blood Types 2 Worksheet where we incorporate the Rh factor on page 186 of this teacher guide.

What did we learn?

1. What are the four parts of blood and the function of each part?
    a.
    b.
    c.
    d.

2. What does your body do to help protect itself if you get cut?

3. Do you have more red or white blood cells?

Taking it further

1. What are some of the dangers of a serious cut?

Name_____ Date_____

🎖 Blood Types 1 Worksheet

On the chart below, the donor blood type is on the top row. Below each blood type, write the blood type to which that donor could safely donate blood.

Blood Donors

A	B	AB	O

On the chart below, the recipient blood type is on the top row. Below each blood type, write the blood type from which that recipient could safely receive blood. Which blood type is the universal donor and which is the universal recipient?

Blood Recipients

A	B	AB	O

The Human Body // 185

Name _____ Date _____

🎗 Blood Types 2 Worksheet

On the chart below, the donor blood type is on the top row. Below each blood type, write the blood type to which that donor could safely donate blood.

Blood Donors

A+	A-	B+	B-	AB+	AB-	O+	O-

On the chart below, the recipient blood type is on the top row. Below each blood type, write the blood type from which that recipient could safely receive blood. Which blood type is the universal donor and which is the universal recipient?

Blood Recipients

A+	A-	B+	B-	AB+	AB-	O+	O-

God's Design: Life | The Human Body | Day 103 | Unit 5 Lesson 27 | Name

27 The Respiratory System
A breath of fresh air

🧪 Supply list

☐ Copy of "The Respiratory System" worksheet

🧪 The Respiratory System Worksheet

Complete the Respiratory System Worksheet on the back of this page. It is based on the diagram on page 227 of your student book.

🎖 Respiration Challenge

What is the most amazing thing you learned about respiration from the information on page 228 of the student book? Be sure to explain it to your teacher!

🧠 What did we learn?

1. Describe the breathing process.

2. How do the circulatory and respiratory systems work together?

3. Where inside the lungs does the exchange of gases occur?

4. What are the major parts of the respiratory system?

🚀 Taking it further

1. How do you suppose your body keeps food from going into your lungs and air from going into your stomach when both enter your body in the back of your throat?

2. How does your respiratory system respond when you exercise?

3. How does your respiratory system respond when you are sleeping?

Name _____   Date _____

🧪 The Respiratory System Worksheet

Use the words below to label the diagram of the respiratory system.

| Alveoli | Diaphragm | Nasal cavity | Throat |
| Bronchial tube | Lung | Nose | Trachea |

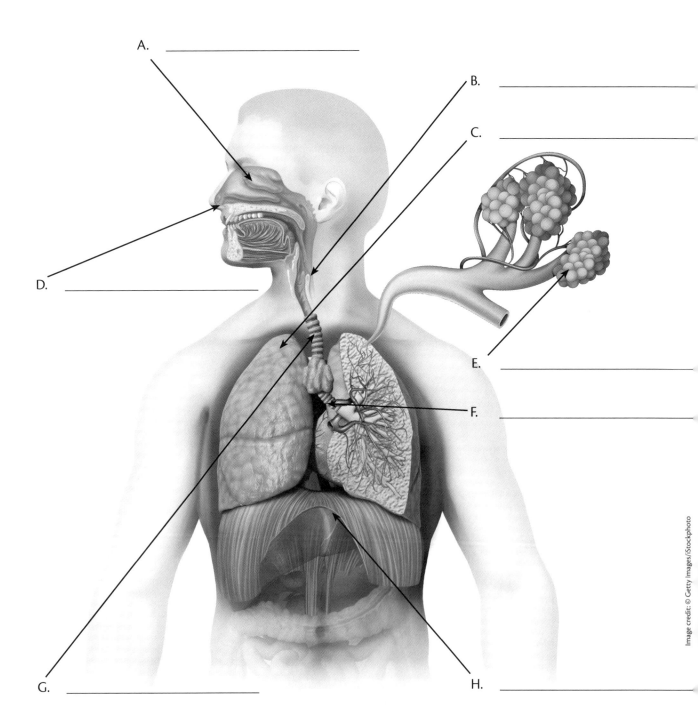

A. _____

B. _____

C. _____

D. _____

E. _____

F. _____

G. _____

H. _____

188 // The Human Body

| God's Design: Life | The Human Body | Day 104 | Unit 5 Lesson 28 | Name |

28 The Lungs

Are there balloons inside my chest?

Supply list

☐ Cloth tape measure

☐ 1 or more balloons

☐ Stopwatch

How Big Are My Lungs?

Want to see how much air your lungs can hold? Follow the instructions for this activity on page 230 of your student book.

Breath Test

Exercise affects our breathing rates, and this activity allows you to explore this fact. Follow the procedure on page 230 of the student book. How does this test of your breathing after exercise relate to the activity you did in lesson 24 with your pulse after exercising?

Word Puzzle Challenge

Use the vocabulary words you learned in this unit to create a word puzzle or word game. Give it to someone — your teacher or another student — to see how much they remember.

What did we learn?

1. How does your body keep harmful particles from entering your lungs?

2. How are your lungs similar to a balloon?

3. How are your lungs different from a balloon?

🚀 Taking it further

1. What can you do to keep your lungs healthy?

2. If you breathe in oxygen and breathe out carbon dioxide, how can you help someone who is not breathing by breathing into his or her lungs when you do CPR?

God's Design: Life | The Human Body | Day 107 | Unit 6 Lesson 29 | Name

The Skin
Keeping your insides in

Supply list
☐ Hand lotion
☐ Mirror

Examining Your Skin

Complete the activity on page 234 of your student book to learn more about the skin on your body. How is the skin different in each area? How is the skin the same in each area?

back of your hand:

fingertips:

bottoms of your feet:

forehead:

knee:

Skin Pigment Challenge

After reading the text on page 235 of the student book, see if you can answer the following questions:

1. What is the most important role of melanin?

2. What color is the pigment carotene?

3. If your body cannot produce melanin, you have what condition?

What did we learn?

1. What are the purposes of skin?

2. How does skin help you stay healthy?

3. How does skin allow you to move without getting stretched out?

🚀 Taking it further

1. Other than skin, in what other ways does your body keep out germs?

2. What skin problems might you experience in a dry climate?

3. What skin problems might you experience in a very moist or humid climate?

4. What are the dangers of a serious burn?

30 Cross-section of Skin

What's below the surface?

Supply list

☐ Notebook paper
☐ Copy of "Skin Word Search"

Skin Word Search

Complete the Skin Word Search puzzle on the back of this worksheet.

Hair Production Challenge

Do some research and find out what causes baldness. Share what you find with your class or family, or you can do an optional activity — using the diagram on page 237 of your student book, recreate this drawing. Be sure to label and color it.

What did we learn?

1. What are the three layers of skin?

2. What is the purpose of the sebaceous gland?

3. In which layer are most receptors located?

Taking it further

1. Explain what happens to your skin when you pick up a pin.

2. How does your skin help regulate your body temperature?

Name _____   Date _____

🧪 Skin Word Search

Find the following words in the puzzle below.

Barrier	Epidermis	Germs	Oil	Skin
Dermis	Fat	Glands	Nerves	Subcutaneous
Elastin	Follicle	Keratin	Protection	Sweat

```
F E P I D E R M I S S I D O I
S K N E D F T F E Q E R D S J
A N I O P R T A L P D F A L G
U E S U B C U T A N E O U S D
S T K F R M S E S W R L K J E
F O I L L M N B T I M L O D R
P O N H H B E R I N I I S E V
S E K E R A T I N G S C A S A
V G C X F R O U U E S L F D R
Y L R I T R T N E R V E S K R
N A T D G I B U G M F K H I L
E N S M I E L K F S W E A T D
R D I W P R O T E C T I O N G
A S A R E W I O P Y M I T T H
G F J K Y T L L T O I Y T E F
```

194 // The Human Body

| God's Design: Life | The Human Body | Day 109 | Unit 6 Lesson 31 | Name |

Fingerprints

You are unique.

🧪 Supply list

☐ Pencil and paper

☐ Clear tape

☐ Index cards

🧪 Fingerprint Identification

Now that you know about the arches, loops, and whorls on your fingertips, collect and examine some other people's fingerprints. Follow the procedure on page 241 of the student book.

🏅 Supplies for Challenge

☐ Light corn syrup   ☐ Newspaper

☐ Red food coloring   ☐ White paper

☐ Paper   ☐ Eyedropper

☐ Meter stick   ☐ Copy of "Blood Splatter Chart"

🏅 Forensic Science

Note: This challenge uses red food coloring — this can stain counters and clothing, so choose your area to do the challenge carefully. It may be best, if weather allows, for it to be done outside.

Using the instructions on page 242 of the student book, complete the Blood Spatter Chart on page 197 of the teacher guide. You will also need a ruler that measures centimeters and millimeters for the challenge.

🧠 What did we learn?

1. Where can friction skin be found on your body?

The Human Body // 195

2. When are fingerprints formed?

3. What are the three major groups of fingerprints?
    a.

    b.

    c.

4. Can you identify identical twins by their fingerprints?

Taking it further

1. What are some circumstances where fingerprints are used?

2. Why do prints only occur on the hands and feet?

3. Do children's fingerprints match their parents' prints?

Name _____   Date _____

🎖 Blood Splatter Chart

Height from which "blood" was dropped	Diameter of splatter
2 cm	
4 cm	
6 cm	
8 cm	
10 cm	
15 cm	
20 cm	
30 cm	
40 cm	
50 cm	
75 cm	
100 cm	

God's Design: Life | The Human Body | Day 111 | Unit 6 Lesson 32 | Name

The Immune System
Keeping you healthy

Supply list
☐ Research materials on allergies, diabetes, and arthritis

How Healthy Are You?
Follow the instructions on page 244 in your student book — you will need several sheets of notebook paper. Fill in the paper as needed over a week. What did you learn? Do you have healthy habits?

Supplies for Challenge
☐ Research materials on vaccines and antibiotics

Helping Our Immune System
You've learned about some amazing things God included in the design of our bodies to keep us healthy. Sometimes we get sick, though, but some amazing medicines help keep us healthy. Do a short written or oral report based on the information you read on page 245 of the student book. Be sure you can explain the difference between vaccines and antibiotics.

What did we learn?

1. What are the major parts of the immune system?

2. What are the two major types of "germs" that make us sick?
   a.

   b.

3. How do tears and mucus help fight germs?

 **Taking it further**

1. Why are a fever and a mosquito bite both indications that your immune system is working?

God's Design: Life | The Human Body | Day 112 | Unit 6 Lesson 33 | Name

33 Genetics
Why you look like you do

Supply list
☐ Copy of "Genetics Quiz"

Genetics Quiz

Follow the instructions on the Genetics Quiz on the back of this worksheet to see what traits you inherited from each of your parents.

Supplies for Challenge

☐ DNA model kit (optional but recommended)
☐ Copy of "DNA Puzzle Pieces" (make additional copies for a longer chain)

DNA Challenge

After reading more about DNA in your student book (pages 247-248), make a model of DNA using the puzzle pieces on page 203 of the teacher guide. If doing the challenge with younger kids, the teacher can cut out the puzzle pieces.

What did we learn?

1. What are genes?

2. Why do children generally look like their parents?

Taking it further

1. If parents look very different from each other what will their children look like?

2. In the past, evolutionists claimed that man evolved independently in different parts of the world and this is where the races came from. If this were true, how likely would it be that the different races could have children together?

The Human Body // 201

Name _____ Date _____

🧪 Genetics Quiz

For each item below, mark which box applies to you, your mom, and your dad to see from whom you inherited each trait. For example, if you can touch your tongue to your nose without using your fingers put a check mark next to the S for self. If your mom or dad can do that put a check mark next to M for Mom and/or D for Dad. If you are adopted you may not know the information below about your birth parents, just mark the S spaces. This will give you an idea of some traits your birth parents may have.

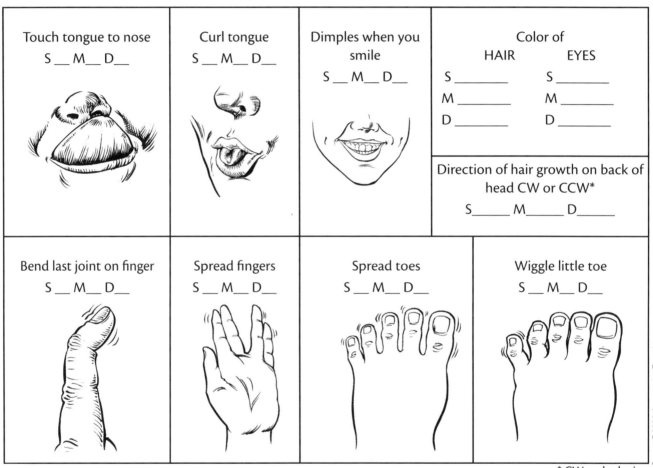

* CW = clockwise
CCW = counter clockwise

Which traits did you inherit from your mother?

Which traits did you inherit from your father?

It might be fun to use this test with your grandparents to try to further trace where these traits came from.

DNA Puzzle Pieces

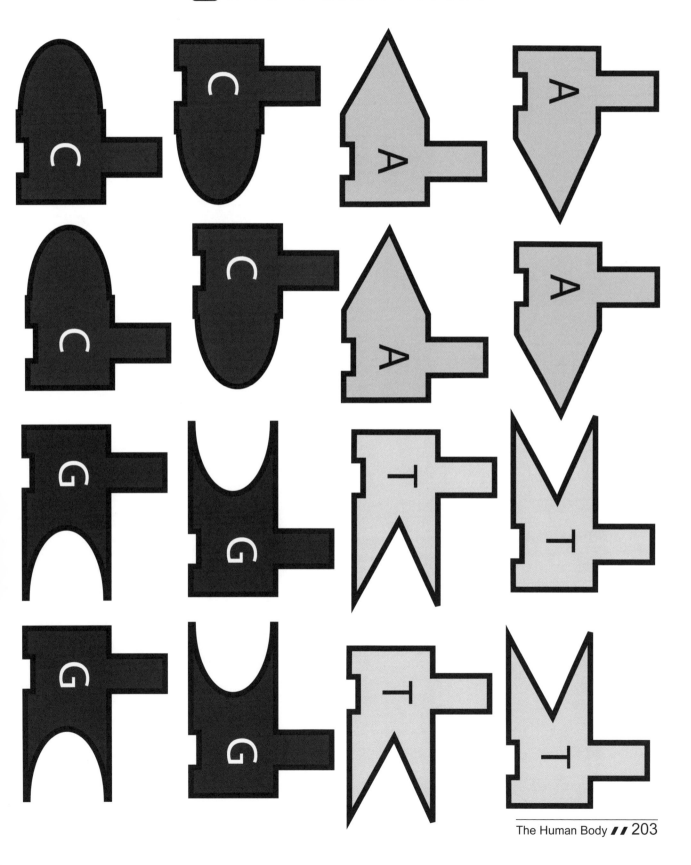

| God's Design: Life | The Human Body | Day 116 | Unit 6 Lesson 34 | Name |

34 Body Poster

Putting it all together

Final Project supply list

☐ Newsprint or other large paper

☐ Markers

☐ Scissors

☐ Tape

☐ Anatomy book

Body Poster

Make a full-size drawing of your body using the instructions on page 252 of your student book. Be sure to explain how each of the systems of your body work after completing the poster.

Body Research Challenge

Choose one of the topics listed on page 252. Research the topic at your local library or online with your teacher's permission. Create a presentation for your class or teacher.

What did we learn?

1. Name the eight body systems you have learned about.

    a.

    b.

    c.

    d.

    e.

    f.

    g.

    h.

2. How do some of the different systems work together?

🚀 Taking it further

1. What other systems can you think of that are in your body but were not discussed in this book?

2. How do you see evidence of God the Creator in the design of the human body?

God's Design: Life | The Human Body | Day 118 | Unit 6 Lesson 35 | Name

35 Conclusion

Appreciating the human body

Supply list

☐ Bible

☐ Paper

☐ Pencil

☐ Concordance (optional)

Appreciating the Human Body

Read the following Bible verses with your teacher. Discuss the kind of relationship God wants to have with each of us.

Matthew 10:29–31

Isaiah 44:24

Psalm 24:1

Psalm 139:13–18

Bonus: Using a concordance, look up verses in the Bible that talk about the body. What does the Bible say about the body? Discuss these verses with your teacher and write a thank you note to God, thanking Him for making you special.

Animal Worksheets

for Use with

The World of Animals

(*God's Design: Life* Series)

| | God's Design: Life | The World of Animals | Day 119 | Unit 1 Lesson 1 | Name |

1 The World of Animals

Is it a mouse or a moose?

 Supply list

☐ **Note:** Throughout this book you may wish to have additional resources with color pictures and information about each group of animals being studied. Animal encyclopedias or other Master Books resources, such as *44 Animals of the Bible, The Complete Zoo Adventure, The Complete Aquarium Adventure*, and the Marvels of Creation series; the Answers in Genesis Zoo Guide and Aquarium Guide, can be valuable in enhancing the lessons.

 Animal Charades

This can be a fun game! Pretend to be an animal and have everyone else guess what animal you are. Whoever guesses the animal correctly gets to be the next animal. Choose animals other than mammals, with which you are most familiar.

 Unusual Animals Challenge

Below is a list of unusual animals. See what you can find out about each of these animals from an animal encyclopedia or other source, and prepare a short report to share with your class or teacher.

1. Pangolin
2. Common snipe
3. Echidna
4. Queen Alexandra's Birdwing
5. Grouper
6. Liver fluke
7. Common whelk

Three of them are shown below. Can you identify them?

What did we learn?

1. What are the two major divisions of animals?
   a.

   b.

2. What are two similarities among all animals?
   a.

   b.

Taking it further

1. When did God create the different animal kinds?

2. How is man different from animals?

| | God's Design: Life | The World of Animals | Day 121 | Unit 1 Lesson 2 | Name |

2 Vertebrates

Does it have a backbone?

Supply list

☐ 3-ring binder

☐ 12 or more dividers with tabs

Animal Notebook

As you work your way through this section of the student book, you will be making a notebook that will include your projects. Follow the instructions for the notebook on page 265 of the student book.

Supplies for Challenge

☐ Drawing materials

Notebook Title Page

Here is a chance to be really creative! Use your artistic, computer, and/or literary skills to create a title page for each section of your animal notebook. If you don't know what kinds of animals belong in some of the sections, look them up in an encyclopedia or on the Internet.

What did we learn?

1. What are the two major divisions of the animal kingdom?

    a.

    b.

2. What characteristics define an animal as a vertebrate?

3. What are the five groups of vertebrates?

   a.

   b.

   c.

   d.

   e.

🚀 Taking it further

1. Think about pictures you have seen of dinosaur skeletons. Do you think dinosaurs were vertebrates or invertebrates? Why do you think that?

| | God's Design: Life | The World of Animals | Day 122 | Unit 1 Lesson 3 | Name |

Mammals
The fuzzy creatures

Supply list

☐ Copy of "Mammals Have Fur" worksheet

☐ Samples of hair from as many mammals as possible

☐ Book showing pictures of different mammals

☐ Suggestions for animal books: *Magnificent Mammals* by Buddy and Kay Davis, and *Kingfisher Illustrated Animal Encyclopedia*.

Mammals Have Fur

Complete the Animals Have Fur Worksheet on page 217 of the teacher guide. Use pictures of mammals to describe the fur for animals that you do not have access to. Although people are set apart from animals, compare a sample of your hair to that of some mammals. Add the worksheet to your animal notebook.

Challenge: Mammal Feet

Think about how each of the following mammals walks. Decide which kind of stance that animal has. It might help if you look at pictures of the animals to see how they stand on their feet.

1. Deer
2. Rabbit
3. Giraffe
4. Wolf
5. Skunk
6. Elephant
7. Opossum
8. Chimpanzee
9. Fox

What did we learn?

1. What five characteristics are common to all mammals?

    a.

    b.

    c.

    d.

    e.

2. Why do mammals have hair?

3. Why is a platypus considered a mammal even though it lays eggs?

Taking it further

1. Name some ways that mammals regulate their body temperature.

2. What are some animals that have hair that helps them hide from their enemies?

Name _____ Date _____

🧪 Mammals Have Fur Worksheet

Animal name	Hair length	Hair color	Parts of body covered	Texture of hair	Sample of hair (If available)
Dog					
Cat					
Pig					
Sheep					
Tiger					
Elephant					
Rabbit					
Armadillo					
Musk-ox					
You (human)					

| | God's Design: Life | The World of Animals | Day 123 | Unit 1 Lesson 4 | Name |

4 Mammals: Large & Small

Armadillo to zebra

Supply list

☐ Drawing paper

☐ Markers, colored pencils, or paint

Investigating Mammals

This is your chance to learn more about a specific mammal of your choice. Follow the instructions on page 271 of the student book to complete this activity and answer the questions. **Note:** There is an additional or optional component for older students.

1. What is this animal's habitat — where does it live?

2. How large does this animal grow to be?

3. What does this animal eat?

4. What enemies does this animal have?

5. How quickly does this animal reproduce? How many offspring does it have? How long is the mother pregnant? How long does the baby stay with the mother?

6. What other interesting things did you find out about your animal?

Ruminants

Read the text on page 272 of your student book. Then draw a diagram of the ruminant digestive system. Be sure to label all the parts. Include this drawing in your animal notebook.

What did we learn?

1. What is the largest land mammal?

2. What is the tallest land mammal?

3. What do bears eat?

🚀 Taking it further

1. What do you think is the most fascinating mammal? Why do you think that?

| God's Design: Life | The World of Animals | Day 124 | Unit 1 Lesson 5 | Name |

5 Monkeys & Apes
Primates

Supply list
☐ Copy of "Mammals Word Search"

Mammals Word Search
Complete the Mammals Word Search on page 223 of the teacher guide. Include this in your animal notebook.

Ape Intelligence Challenge
Read the text on page 275 of the student book. It points out an important difference in the abilities of primates and humans. Explain what you learned to your teacher. Be sure to explain why animals can be intelligent but not have language with a grammar structure.

What did we learn?

1. What are two common characteristics of all primates?
   a.
   b.

2. What are the three groups of primates?
   a.
   b.
   c.

3. What is one difference between apes and monkeys?

4. Where do New World Monkeys live?

5. Where do Old World Monkeys live?

6. What is a prehensile tail?

🚀 Taking it further

1. If a monkey lives in South America is it likely to have a prehensile tail?

2. Are you more likely to find a monkey or an ape in a tree in the rainforest?

3. Why do most prosimians have very large eyes?

Name _____ Date _____

🧪 Mammals Word Search

Find the following words in the puzzle below. Words may be horizontal, vertical, or diagonal.

Ape	Fur	Live birth	Monkey	Warm-blooded
Bat	Giraffe	Lungs	Mouse	Whale
Camel	Lemur	Mammary gland	Primate	Zebra

```
D G E U T A F R O P K L H Y U
B E N D R M L J I O S C X F A
A Z E B R A Z C V B C A B S R
M N I D S M F V G F T T J A E
S R E I H M U X A A I M U P U
C O L S W A R M B L O O D E D
C A M E L R H O P T W U P E A
K E Y M O Y U S P Z E S R R W
F R L U G G D R W R G E K H U
M L U L E L I V E B I R T H S
O U E N A A T G L A R M D O B
C N H M O N K E Y V A I A N P
F G Y B U D L E W C F A N T L
H S A L E R U P Q Z F L U H E
R A J C J K W H A L E E S O P
```

The World of Animals **//** 223

| God's Design: Life | The World of Animals | Day 127 | Unit 1 Lesson 6 | Name |

6 Aquatic Mammals

They live in the water?

Supply list

☐ Toothbrush

☐ Stopwatch

☐ 2 cups

☐ Water

☐ Chopped nuts, fruits, or vegetables

Acting Like a Whale

Here are two interesting activities for you to do! How long could you hold your breath for Activity 1? Do you understand how Activity 2 explains how baleen whales get food?

Amazing Whales Challenge

Read the text on page 281 of your student book and learn more about whales. Several types are discussed. Whales are truly designed by God to be well suited for their environment. See what other interesting things you can find out about whales that demonstrate God's special design.

What did we learn?

1. Why are dolphins and whales considered mammals and not fish?

2. What is the main difference between the tails of fish and the tails of aquatic mammals?

3. What is another name for a manatee?

4. Why are manatees sometimes called this?

🚀 Taking it further

1. How has God specially designed aquatic mammals for breathing air?

2. What do you think might be one of the first things a mother whale or dolphin must teach a newborn baby?

God's Design: Life | The World of Animals | Day 128 | Unit 1 Lesson 7 | Name

Marsupials
Pouched animals

🧪 Supply list

☐ Plastic zipper bag ☐ Scissors
☐ Glue ☐ Fake fur or felt
☐ Tagboard or cardboard ☐ Construction paper

🧪 Making a Pouch

Make your own marsupial pouch! Use the instructions on page 284 of the student book to complete the activity.

🎖 Supplies for Challenge

☐ Copy of "Koala Fun Facts" worksheet ☐ Book or Internet sites on koalas

🎖 Koalas Challenge

Research and learn more about these cuddly-looking creatures! See what you can find on koalas in an animal encyclopedia or from an online source. Fill out the Koala Fun Facts Worksheet on the next page.

🧠 What did we learn?

1. What is a marsupial?

2. Name at least three marsupials.

3. How has God designed the kangaroo for jumping?

🚀 Taking it further

1. About half of a kangaroo's body weight is from muscle. This is nearly twice as much as in most animals its size. How might this fact contribute to its ability to hop?

2. How do you think a joey kangaroo keeps from falling out of its mother's pouch when she hops?

The World of Animals ◢◢ 227

Name _____   Date_____

🎖 Koala Fun Facts Worksheet

1. We often hear a koala referred to as a koala bear, but it is not a bear. List three ways that a koala is different from a bear.

   _____

   _____

   _____

2. What is unusual about the koala's pouch? _____

3. What does a baby koala eat? _____

4. What does an adult koala eat? _____

5. What special design features make koalas able to eat this kind of food? _____

   _____

6. How long does a baby koala spend in its mother's pouch? _____

7. How does the mother koala carry her youngster after it leaves the pouch? _____

   _____

8. What is special about the skin on the koala's feet that helps it to climb trees? _____

   _____

9. What is special about the koala's hands that help it survive? _____

   _____

10. How much does a koala sleep? _____

11. About how much does an adult koala weigh? _____

12. What is the average life span of a koala? _____

| | God's Design: Life | The World of Animals | Day 131 | Unit 2 Lesson 8 | Name |

Birds
Fine feathered friends

 Supply list

☐ Copy of "Bird Beaks" and "Bird Feet" worksheets

☐ Bird guides or encyclopedias

Bird Beaks and Bird Feet

Follow the instructions for this activity on page 288 of the student book. Use the worksheets on page 231 and 232 in this teacher guide. There are optional activities you can do as well.

Supplies for Challenge

☐ Animal or bird encyclopedias

Challenge: Birds vs. Reptiles

List some characteristics that may vary among a species due to natural selection. Look through an animal encyclopedia to see examples of these characteristics. Notice that not one of these various characteristics has resulted in a new species. Read the text on page 289 of the student book.

What did we learn?

1. How do birds differ from mammals?

2. How are birds the same as mammals?

🚀 Taking it further

1. How can you identify one bird from another?

2. What birds can you identify near your home?

3. Why might you see different birds near your home in the summer than in the winter?

Name _____   Date _____

🧪 Bird Beaks Worksheet

In the space below, draw the different types of beaks that birds have.

For eating nectar	For eating fish

For eating prey	For eating bugs

For eating nuts and seeds	

The World of Animals // 231

Name _____   Date _____

🧪 Bird Feet Worksheet

In the space below, draw the different types of feet that birds have.

For swimming	For perching in trees

For catching prey	For walking or running

| God's Design: Life | The World of Animals | Day 133 | Unit 2 Lesson 9 | Name |

9 Flight

How do those birds do that?

Supply list

☐ Copy of "God Designed Birds to Fly" worksheet

☐ 1 or more bird feathers

☐ Magnifying glass

Examining a Bird's Feather

Learn about preening and more when you do this simple activity as directed on page 293 of the student book. **Note**: Check your yard for possible feathers to examine. You can purchase a small package of feathers at craft stores.

God Designed Birds to Fly

Complete the God Designed Birds to Fly Worksheet on page 235 of the teacher guide. Follow the additional instructions for this activity on page 293 of the student book. Don't forget to add it to your animal notebook!

Flightless Birds Challenge

Read the text on pages 293 and 294 in the student book. Then be prepared to explain why the wings of flightless birds don't help them fly, but they are not useless because they serve other functions. Give your teacher some examples.

What did we learn?

1. What are some ways birds are designed for flight?

2. What are the three kinds of bird feathers?

    a.

    b.

    c.

3. How does a bird repair a feather that is pulled open?

4. How does a bird's tail work like a rudder?

🚀 Taking it further

1. Why can't man fly by strapping wings to his arms?

2. How do you think birds use their feathers to stay warm?

3. How is an airplane wing like a bird's wing?

Name _____   Date _____

🧪 God Designed Birds to Fly Worksheet

Birds' wings are designed like airfoils to achieve lift. Draw an airfoil.

Birds have three sets of flight feathers (primary, secondary, and tertiary) on their wings in order to change the shape of the wing during flight. Label these feathers on the bird's wing.

Feathers are designed to be strong, light, and flexible. They have a hook and barb design to help them keep their shape. Draw the structure of a feather.

| God's Design: Life | The World of Animals | Day 134 | Unit 2 Lesson 10 | Name |

10 The Bird's Digestive System
They sure eat a lot.

Supply list

☐ Copy of "God Designed the Bird's Digestive System" worksheet

☐ Owl pellet (optional)

Digestive System Worksheet

Draw and label the parts of the bird's digestive system on the God Designed the Bird's Digestive System Worksheet on page 239 of the teacher guide. Use the image on page 295 in your student book as a reference if needed. This worksheet can be added to your animal notebook when you complete it.

Dissect an Owl Pellet

Note: This is an optional exercise. You can purchase owl pellets and complete this activity if you wish. More information is available on page 296 of the student book.

Respiratory System Challenge

Read the text on page 297 of the student book. Answer the following questions:

1. Why did God design a special respiratory system for birds?

2. Do the lungs of birds expand and contract?

Bonus! Make a poster explaining the unique respiratory system of birds, or recreate the diagram on page 297 of the student book.

What did we learn?

1. How does a bird "chew" its food without teeth?

2. What purposes does the crop serve?

Taking it further

1. How is a bird's digestive system different from a human digestive system?

2. How does a bird's digestive system help it to be a better flyer?

Name _____ Date _____

🧪 God Designed the Bird's Digestive System

Birds do not have teeth, so God designed them with a special digestive system. Draw and label the parts of the bird's digestive system.

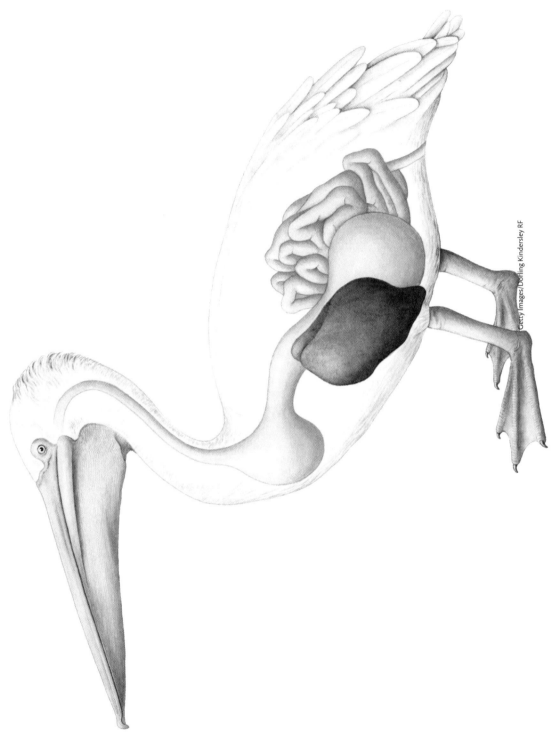

| God's Design: Life | The World of Animals | Day 136 | Unit 2 Lesson 11 | Name |

Fish

Do fish go to school?

Supply list

☐ Goldfish snack crackers
☐ Paper
☐ Glue
☐ Colored pencils

Name a Group Name

Have one person name an animal, then have another name what a group of that type of animal would be called. Here are a few to get you started:

A school of fish, a flock of birds, a gaggle of geese, a herd of elephants, a pride of lions, a pack of wild dogs, a flock of sheep, a brood of vipers, a swarm of flies. For more unusual names for groups of animals (also called collective nouns), see pages 243 and 244 of this teacher guide.

Bonus: See if you can take this list and create a game with it. Even a simple guessing game would be fun!

Fish School

Follow the directions on page 299 of the student book. Be sure to add details to your underwater picture! And take a photo of it to include in your animal notebook.

Challenge: Designed for Speed

Read the text on page 300 of your student book — and then describe to your teacher the difference in placoid scales and leptoid scales.

What did we learn?

1. What makes fish different from other animals?

2. How do fish breathe?

3. Why do some sharks have to stay in motion?

4. What is the difference between warm-blooded and cold-blooded animals?

Taking it further

1. Other than how they breathe, how are dolphins different from fish?

2. How are dolphins like fish?

Animal Group Names

Here are some common and unusual names for groups of animals. Many have more than one, and you may be familiar with some. Others will be a complete surprise! This would also be good to put in your animal notebook.

A parliament of owls

Creature	Collective noun
albatross	rookery
antelopes	herd
ants	army
apes	shrewdness
baboons	troop
bats	colony
bears	sleuth or sloth
bees	swarm
cats	clowder
cattle	mob
cheetahs	coalition
chickens	clutch
coyotes	pack
crocodiles	bask
crows	murder
dogs	pack
dolphins	pod
ducks	paddling
eagles	convocation
eels	bed

Creature	Collective noun
elephants	herd
elk	gang
fish	school
flamingoes	flamboyance
frogs	army
geese	gaggle
giraffes	tower
gnats	cloud
goats	flock
goldfinches	charm
gorillas	band
hawks	kettle
hens	brood
hippopotamuses	bloat
hogs	parcel
hyenas	cackle
jellyfish	fluther
kangaroos	mob
kittens	kindle
lions	pride
mice	nest

A flamboyance of flamingoes

A dazzle of zebras

Creature	Collective noun
sheep	drove
spiders	cluster or clutter
squirrels	scurry
storks	mustering
swans	bank
tigers	ambush
turkeys	rafter
turtles	nest
vipers	nest
vultures	committee
walruses	herd
whales	mob or pod
wolves	pack
woodpeckers	descent
zebras	dazzle

Creature	Collective noun
minnows	shoal
moles	labor
monkeys	barrel
owls	parliament
parrots	company
parrots	pandemonium
peacocks	pride
pelicans	pod
penguins	colony
quail	covey
rabbits	nest
rabbits	herd
rabbits	litter
raccoons	gaze
rats	mischief
ravens	unkindness
rhinoceroses	crash
seagull	squabble
seals	pod

https://en.wikipedia.org/wiki/List_of_English_terms_of_venery,_by_animal

A tower of giraffes

244 *//* The World of Animals

| | God's Design: Life | The World of Animals | Day 137 | Unit 2 Lesson 12 | Name |

Fins & Other Fish Anatomy

Designed for efficiency

Supply list

☐ Copy of "Fish Fins" worksheet

☐ Construction paper

☐ Scissors

☐ Glue

Fish Fins

Cut fins from construction paper and glue them in the correct places on the Fish Fins Worksheet on page 247 of this teacher guide. You can use the image on page 302 of the student book as a reference. Label the fins and color the fish. This page can be included in your animal notebook.

Challenge: Nervous System

Read the text on page 303 of the student book, then give a short presentation of the nervous system of a fish. Be sure to say what was most amazing in all you learned!

What did we learn?

1. What is the purpose of a swim bladder?

2. How did God design the fish to be such a good swimmer?

🚀 Taking it further

1. How does mucus make a fish a more efficient swimmer?

2. How has man used the idea of a swim bladder in his inventions?

3. What other function can fins have besides helping with swimming?

4. What similar design did God give to both fish and birds to help them get where they are going?

Name _____ Date _____

🔬 Fish Fins Worksheet

The fish below is missing its fins. Cut fins from construction paper and glue them to the fish to make it a good swimmer. The list below should help you know what fins to add.

Pectoral fins   Pelvic fins   Dorsal fin   Anal fin   Caudal fin

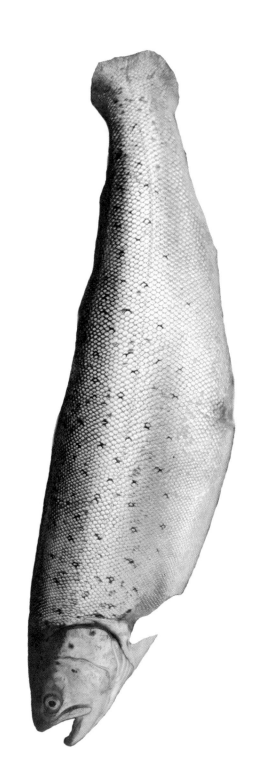

Getty Images/iStockphoto

The World of Animals // 247

| | God's Design: Life | The World of Animals | Day 138 | Unit 2 Lesson 13 | Name |

13 Cartilaginous Fish
No bones about it!

🧪 Supply list

☐ Modeling clay

🧪 Clay Models

Use modeling clay to make models of cartilaginous fish such as sharks and rays. When you are done, take pictures to put in your animal notebook.

🏅 Supplies for Challenge

☐ Pictures of cartilaginous fish (includes sharks, kites, rays)

🏅 Manta Rays Challenge

Draw an image of a manta ray. Then include a short written report of what you learned about manta rays from the text you read.

🧠 What did we learn?

1. How do cartilaginous fish differ from bony fish?

2. Why is a lamprey called a parasite?

3. Why can sharks and stingrays be dangerous to humans?

🚀 Taking it further

1. Why are shark babies born independent?

2. What do you think is the shark's biggest natural enemy?

| God's Design: Life | The World of Animals | Day 141 | Unit 3 Lesson 14 | Name

14 Amphibians
Air or water?

 No Supplies Needed

 Warm-blooded/Cold-blooded

Read the activity instructions carefully on page 309 of your textbook and see if you can act out the part of a warm-blooded or cold-blood animal or both. What did you learn from the activity?

Animal Communication

Option 1: Research other ways that animals communicate with each other. Look for patterns on their bodies, listen to the noises they make, and even smell the scents they leave behind. These are all ways that animals talk to each other.

Option 2: Try to practice communicating the way frogs do. Have one person make a short noise every second. Have a second person make a different noise in between the first person's noises, then try to carry on a conversation with a third person. The trick is that you and your partner can only talk when both of the other people are quiet. This is very difficult to do. Does this help you appreciate how well God designed frogs to communicate with each other?

What did we learn?

1. What are the characteristics that make amphibians unique?

2. How can you tell a frog from a toad?

3. How can you tell a salamander from a lizard?

🚀 Taking it further

1. What advantages do cold-blooded animals have over warm-blooded animals?

2. What advantages do warm-blooded animals have over cold-blooded animals?

3. Why are most people unfamiliar with caecilians?

| God's Design: Life | The World of Animals | Day 142 | Unit 3 Lesson 15 | Name |

15 Amphibian Metamorphosis
Making a change

Supply list

☐ Copy of "Amphibian Life Cycle" worksheet

☐ Tadpoles (optional)

☐ Tank

☐ Food for raising a frog

 Amphibian Life Cycle

Complete the Amphibian Life Cycle Worksheet on page 255 of the teacher guide using the instructions found on page 312 of the Student Book.

Optional: As noted in the textbook, the Grow a Frog activity is an optional one or could be done for bonus points if the teacher assigns it.

Unusual Amphibians

Read the text on page 313 of the student book. Look at an animal encyclopedia or other source to find out about other unusual methods of amphibian reproduction. You can choose one unusual amphibian and tell your teacher about how that amphibian reproduces.

What did we learn?

1. Describe the stages an amphibian goes through in its life cycle.

2. What are gills?

3. What are lungs?

 Taking it further

1. Does the amphibian lifecycle represent molecules-to-man evolution? Why or why not?

Name _____   Date _____

Amphibian Life Cycle Worksheet

Draw pictures representing each stage of an amphibian's life cycle.

Egg	Tadpole (larva)

Metamorphosis	Adult

The World of Animals **255**

| | God's Design: Life | The World of Animals | Day 143 | Unit 3 Lesson 16 | Name |

16 Reptiles
Scaly animals

 Supply list

☐ Paper

☐ Pictures of reptiles

☐ Sequins or flat beads

☐ Glue

Scaly Picture

Draw an outline of a reptile and then glue sequins wherever the creature would have scales on its body. Turtles only have scales on their legs, feet, tail, neck, and head. Alligators have very large scales on their bodies. Snakes have scales over their entire bodies. Include this picture in your animal notebook.

Dinosaurs Challenge

Read more about dinosaurs on pages 315 and 316 of the student book. Choose a particular dinosaur and research it, or learn more at answersingenesis.org about the controversy over whether dinosaurs were warm-blooded or cold-blooded. You can get creative and design a page for your animal notebook on dinosaurs — you can use magazine clippings, drawings you make, facts you find out, etc.

What did we learn?

1. What makes reptiles different from amphibians?

2. What are the four groups of reptiles?

    a.

    b.

    c.

    d.

🚀 Taking it further

1. How do reptiles keep from overheating?

2. What would a reptile likely do if you dug it out of its winter hibernation spot?

| God's Design: Life | The World of Animals | Day 146 | Unit 3 Lesson 17 | Name |

17 Snakes
Those hissing, slithering creatures

Supply list
☐ An open area on the floor for moving about

Slithering Like a Snake

Get down on the floor and try to move like the instructions tell you on page 321 of the student book. It may not be easy to match the movements of a snake, but you will have fun trying!

Snake Research

There are over 2,300 species of snakes. We have only scratched the surface in this lesson. Use an animal encyclopedia or the Internet to find out more about snakes. Make a snake presentation to include in your animal notebook. Include pictures and interesting facts that you find about various kinds of snakes.

What did we learn?

1. How are snakes different from other reptiles?

2. What are the three groups of snakes?

3. How is a snake's sense of smell different from that of most other animals?

4. What is unique about how a snake eats?

🚀 Taking it further

1. How are small snakes different from worms?

2. If you see a snake in your yard, how do you know if it is dangerous?

| God's Design: Life | The World of Animals | Day 148 | Unit 3 Lesson 18 | Name |

Lizards

Chameleons and Gila monsters

🧪 Supply list

☐ Paper

☐ Face paint

☐ Markers, colored pencils or paints

🧪 Animal Camouflage

Follow the instructions on page 324 of your textbook to create a "camouflaged" image of a chameleon. Be creative with your background image! Include it in your notebook.

🧪 People Camouflage

Read the instructions on page 324 of the student book. Don't forget to take a photo of yourself!

🏅 Large Lizards

Read the information on page 325 of the student book. Do some research and see what you can find out about Komodo dragons. Make a Komodo dragon page for your animal notebook.

🧠 What did we learn?

1. List three ways a lizard might protect itself from a predator.

    a.

    b.

    c.

2. What do lizards eat?

🚀 Taking it further

1. Horny lizards are short compared to many other lizards and are often called horny toads. What distinguishes a lizard from a toad?

2. Why might some people like having lizards around?

3. How does changing color protect a lizard?

4. What other reasons might cause a lizard to change colors?

| God's Design: Life | The World of Animals | Day 149 | Unit 3 Lesson 19 | Name |

Turtles & Crocodiles

Turtle or tortoise, crocodile or alligator — how do you tell?

Supply list

☐ Copy of "How Can You Tell Them Apart?" worksheet

☐ Tape (cloth or first aid tape is best)

☐ Sink

How Can You Tell Them Apart?

Complete the How Can You Tell Them Apart? Worksheet on page 265 of this teacher guide by drawing pictures showing how to tell turtles from tortoises and crocodiles from alligators. This worksheet can be included in your animal notebook.

Turtle Feet and Tortoise Feet

Complete the activity on page 328 of the student book. What did you learn about the difference in turtle feet and tortoise feet?

Turtle Shells Challenge

Read the text on page 328 of the student book. Were you surprised to learn about the special design of turtle shells? What amazed you the most?

What did we learn?

1. Where do turtles usually live?

2. Where do tortoises usually live?

3. How does the mother crocodile carry her eggs to the water?

4. Why can't you take a turtle out of its shell?

5. How do crocodiles stalk their prey?

🚀 Taking it further

1. Why might it be difficult to see a crocodile?

Name _____   Date _____

🧪 How Can You Tell Them Apart?

How can you tell if you are looking at a turtle or a tortoise?

1. Look at its habitat. Turtles spend most of their time in the water and tortoises live on the land.

2. Look at its feet and legs. Turtles have webbed feet or paddle-like legs. Tortoises have short, stocky legs with clawed feet. Draw pictures showing the difference below.

Turtle feet	Tortoise feet

How can you tell if you are looking at a crocodile or an alligator?

1. Look at the shape of its snout. Crocodiles have long, thin snouts. Alligators have shorter, more rounded U-shaped snouts.

2. Look at its mouth from the side when it is closed. Can you see any teeth sticking out? If you cannot see any teeth, you are looking at an alligator. If you can see one or more teeth sticking out on the bottom, you are looking at a crocodile. Draw pictures showing the differences below.

Top view of snouts

Crocodile	Alligator

Side view of snouts

Crocodile	Alligator

| God's Design: Life | The World of Animals | Day 152 | Unit 4 Lesson 20 | Name |

20 Invertebrates

Creatures without a backbone

🧪 Supply list

☐ A good imagination

☐ White board with markers (optional)

🧪 Invertebrate Repeat Game

Enjoy a simple memory game by following the instructions on page 331 of the student book.

🧪 Invertebrate Pictionary

Note: This is an optional activity. Follow the instructions on page 331, instead of the repeat game activity.

🏅 Challenge: Symmetry

Read the text on pages 331 and 332 about symmetry of animals. Recreate one of the examples in the text, using a ruler to make the lines. Or, take an image of an animal you choose and try to determine what type of symmetry, if any, it has.

🧠 What did we learn?

1. What are some differences between vertebrates and invertebrates?

2. What are the six categories of invertebrates?
    a.
    b.
    c.
    d.
    e.
    f.

🚀 Taking it further

1. Why might we think that there are more vertebrates than invertebrates in the world?

| | God's Design: Life | The World of Animals | Day 153 | Unit 4 Lesson 21 | Name |

21 Arthropods
Invertebrates with jointed feet

🧪 Supply list
☐ Copy of the "Arthropod Pie Chart"

🧪 Arthropod Pie Chart

Complete the Arthropod Pie Chart on page 271, using the instructions found on page 334 of your student book. When complete, add it to your notebook.

🏅 Supplies for Challenge
☐ Balloons
☐ Newspaper
☐ String
☐ Flour

🏅 Exoskeleton Model Challenge

Create an exoskeleton model by following the procedure on page 335 of the student book.

🧠 What did we learn?

1. What do all arthropods have in common?

2. What is the largest group of arthropods?

🚀 Taking it further

1. How are endoskeletons (internal) and exoskeletons (external) similar?

2. How are endoskeletons and exoskeletons different?

3. Why should you be cautious when hunting for arthropods?

Name _____   Date _____

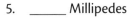 Arthropod Pie Chart

Match the arthropod groups below to the chart sections, showing the relative size of each group.

1. _____ Insects
2. _____ Crustaceans
3. _____ Arachnids
4. _____ Centipedes
5. _____ Millipedes

| God's Design: Life | The World of Animals | Day 154 | Unit 4 Lesson 22 | Name |

22 Insects

Don't let them bug you.

Supply list

☐ Copy of "Water Skipper Pattern"
☐ Paint
☐ Bowl of water
☐ 3 Styrofoam balls per child
☐ 4 pipe cleaners per child

☐ 1 index card per child
☐ Scissors
☐ 2 toothpicks per child
☐ Paper
☐ Tape

Insect models

Discuss each part of the insect as you complete the insect model. Instructions are found on page 337 of the student book. Take pictures of the models and include them in your animal notebook.

Water Skipper Model

Follow the instructions on page 337. On page 275 of the teacher guide, you will find the Water Skipper Pattern. Did the card float?

Optional Activity

You can play the game "Cootie" from Milton Bradley as you review the parts of the insect. You can do this instead of the other activities for this lesson if your teacher allows it.

Insect Anatomy Challenge

Read the text on page 338 of the student book. You can tell your teacher what you learned or demonstrate your understanding in another way. You can make a drawing of an insect, labeling the parts, or focus on one part of the insect anatomy you learned about and do a short report on it.

What did we learn?

1. What characteristics classify an animal as an insect?

2. How can insects be harmful to humans?

3. How can insects be helpful?

Taking it further

1. How might insects make noise?

Name _____ Date _____

 # Water Skipper Pattern

Folded edge

The World of Animals 275

| God's Design: Life | The World of Animals | Day 156 | Unit 4 Lesson 23 | Name |

23 Insect Metamorphosis
Making a change

Supply list
☐ Copy of "Stages of Incomplete Metamorphosis" and "Stages of Complete Metamorphosis" worksheets
☐ Butterfly larvae and habitat (optional)

Metamorphosis Worksheet
Complete the Stages of Metamorphosis Worksheet on page 279 of this teacher guide. Add to your notebook.

Observe metamorphosis
Note: This is a long-term project that the teacher may want to assign if the student is interested or for bonus points. Learn more about it on page 341 of the textbook.

Fireflies Challenge
Read the text on page 341 of the student book. Using the image on the page, make your own drawing of a firefly or create an informational poster about fireflies using details that you learned.

What did we learn?
1. What are the three stages of incomplete metamorphosis?
   a.

   b.

   c.

2. What are the four stages of complete metamorphosis?

   a.

   b.

   c.

   d.

🚀 Taking it further

1. What must an adult insect look for when trying to find a place to lay her eggs?

Name _____   Date _____

🧪 Stages of Incomplete Metamorphosis

Draw and color the different stages of incomplete metamorphosis.

Egg	Nymph	Adult

🧪 Stages of Complete Metamorphosis

Draw and color the different stages of complete metamorphosis.

Egg	Larva
Pupa	Adult

God's Design: Life | The World of Animals | Day 157 | Unit 4 Lesson 24 | Name

24 Arachnids
Spiders and such

🧪 Supply list

Required activity:

☐ Several large and small marshmallows for each child

☐ Flexible wire (about 4 inches for each child)

☐ 1 toothpick per child

☐ 4–6 pipe cleaners per child

🧪 Spider and Scorpion Models

Follow the instructions on page 343 of the student book. There are two optional activities you can do as well.

First optional activity:

☐ Peanut butter

☐ 2 round crackers per child

☐ 2 raisins per child

☐ 8 pretzel sticks per child

Second optional activity:

☐ Spider webs

☐ Powdered sugar

☐ Magnifying glass

🏅 Tarantulas Challenge

Tell your teacher what you learned about these creatures after reading the text on page 344.

What did we learn?

1. How do arachnids differ from insects?

2. Why are ticks and mites called parasites?

Taking it further

1. Why don't spiders get caught in their own webs?

| | God's Design: Life | The World of Animals | Day 158 | Unit 4 Lesson 25 | Name |

25 Crustaceans
Are they crusty?

Supply list

☐ Modeling clay

☐ Magnifying glass for optional activity

Clay Models

Make clay models of several different crustaceans. Note the similarities and differences between the animals. Take pictures of the models to add to your animal notebook.

Woodlouse Exploration

Note: This is an optional exercise. Follow the instructions on page 346 of the student book for clues on where to find a woodlouse and a picture of one.

Supplies for Challenge

☐ Research materials on crustaceans

Crustacean Research Challenge

Do some research in an animal encyclopedia, on the Internet, or other source and find out what you can about the animals listed below. Create one or more pages in your animal notebook using the information you find on the following animals:

Krill

Plankton

Cleaner Shrimp

Barnacles

Water slater

The World of Animals 283

What did we learn?

1. What do all crustaceans have in common?

2. What are some ways that the crayfish is specially designed for its environment?

Taking it further

1. Why might darting backward be a good defense for the crayfish?

2. At first glance, scorpions and crayfish (or crawdads) look a lot alike. How does a scorpion differ from a crayfish?

3. How can something as large as a blue whale survive by eating only tiny crustaceans?

4. If you want to observe crustaceans, what equipment might you need?

| | God's Design: Life | The World of Animals | Day 159 | Unit 4 Lesson 26 | Name |

26 Myriapods

How many shoes would a centipede have to buy?

Supply list

☐ A good memory and a sense of adventure
  (A baseball cap might be fun, too.)

 Arthropod Baseball

Learn more about arthropods as you enjoy this fun game! Learn the rules on page 348 of your student book.

Supplies for Challenge

☐ Modeling clay
☐ Pipe cleaners or craft wire

Myriapod Models

Demonstrate your understanding of the differences between centipedes and millipedes by making models of each. More information and instructions are found on page 349 of the student book.

What did we learn?

1. How can you tell a centipede from a millipede?

2. What are the five groups of arthropods?

3. What do all arthropods have in common?

🚀 Taking it further

1. What are some common places you might find arthropods?

2. Arthropods are supposed to live outside, but sometimes they get into our homes. What arthropods have you seen in your home?

| God's Design: Life | The World of Animals | Day 162 | Unit 5 Lesson 27 | Name |

Mollusks

Creatures with shells

 Supply list

☐ Several sea shells

☐ Shell identification guide

Shell Identification

Follow the instructions on page 352 of the student book. Try to use a variety of shells if possible. Take pictures to include in your notebook or do a drawing of the shells instead.

Supplies for Challenge

☐ Balloon

The Nautilus

Learn about the nautilus by reading the text on page 353 of the student book. Do the activity to learn more about the propulsion system of these creatures.

What did we learn?

1. What are three groups of mollusks?

2. What body structures do all mollusks have?

3. How can you use a shell to help identify an animal?

🚀 Taking it further

1. How are pearls formed?

| God's Design: Life | The World of Animals | Day 163 | Unit 5 Lesson 28 | Name |

28 Cnidarians

Jellyfish, coral, and sea anemones

Supply list

☐ Copy of "Coral Pattern" worksheet

☐ Baking dish

☐ Thin cardboard

☐ Ammonia

☐ Salt

☐ Food coloring

☐ Liquid bluing (located with laundry soap in grocery store)

Grow Your Own Coral Colony

Follow the instructions on page 356 of the student book. The Coral Pattern Worksheet is on page 291 of the teacher guide. **Note:** This activity uses both food coloring and ammonia; students should have adult supervision.

Supplies for Challenge

☐ Live hydra specimens (optional—can be ordered from a science supply store)

Man O' War Challenge

Note: It is optional if you wish to use live hydra specimens that you order. Instead you can do a short report on what you learned from the information on page 357 or a research project on how hydras reproduce.

What did we learn?

1. What characteristics do all cnidarians share?

2. What are the three most common cnidarians?

🚀 Taking it further

1. How do you think some creatures are able to live closely with jellyfish?

2. Why do you think an adult jellyfish is called a medusa?

3. Jellyfish and coral sometimes have symbiotic relationships with other creatures. What other symbiotic relationships can you name?

Name _____   Date _____

🧪 Coral Pattern Worksheet

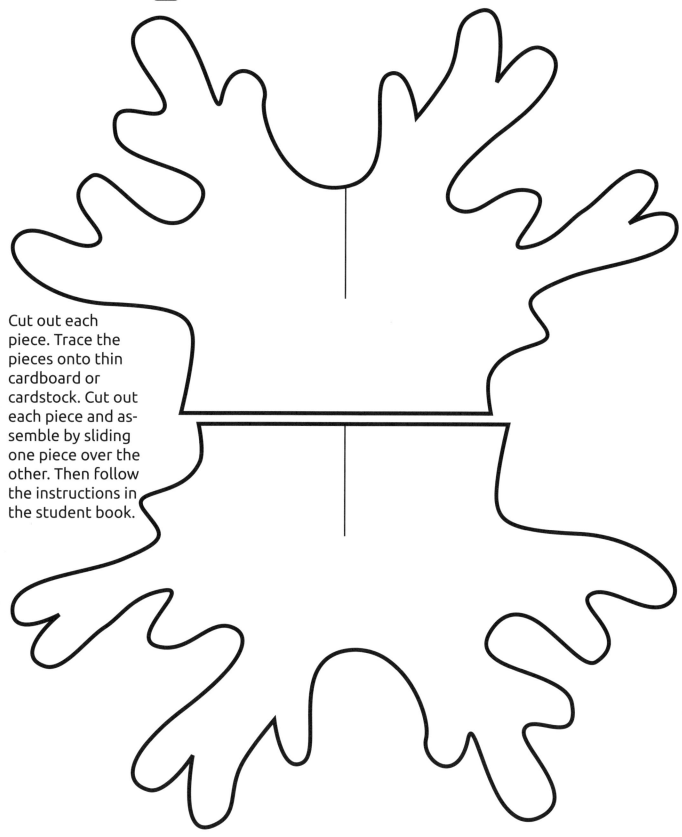

Cut out each piece. Trace the pieces onto thin cardboard or cardstock. Cut out each piece and assemble by sliding one piece over the other. Then follow the instructions in the student book.

| God's Design: Life | The World of Animals | Day 164 | Unit 5 Lesson 29 | Name

 Echinoderms
Spiny-skinned creatures

 Supply list

☐ Salt dough (1 cup salt, 1 cup flour, water to make a stiff dough)

☐ Tagboard or cardboard

☐ Mini chocolate chips

☐ If possible: a real (dead) starfish or sand dollar (sometimes available at craft stores)

Starfish Model

Use the instructions on page 359 of the student book to complete a starfish model.

Observing Echinoderms

Note: This is an optional activity. For more information, see the details on page 360 of your student book.

Supplies for Challenge

☐ Optional: A preserved starfish for dissection

☐ Plastic gloves

☐ Dissecting tray and scalpel

Water Vascular System Challenge

Order dissecting materials from a science supply store and follow the instructions that come with the starfish specimen. Or, if you prefer just to watch a dissection, you can find several sites on the Internet that show step-by-step photos of an actual starfish dissection.

CAUTION: The teacher should supervise use of the dissection tools if the activity is done this way. If a video is being substituted, it is important that the teacher prescreen the video to make sure it is age appropriate for the student.

Draw a diagram of the starfish water vascular system (see diagram on page 360 in the student book). Label all of the parts and include in your animal notebook.

What did we learn?

1. What are three common echinoderms?

    a.

    b.

    c.

2. What do echinoderms have in common?

Taking it further

1. Why would oyster and clam fishermen not want starfish in their oyster and clam beds?

2. What would happen if the fishermen caught and cut up the starfish and then threw them back?

3. What purpose might the spikes serve on echinoderms?

God's Design: Life | The World of Animals | Day 166 | Unit 5 Lesson 30 | Name

Sponges

How much water can a sponge hold?

 Supply list

☐ Synthetic sponges (and if possible, a real sea sponge)
☐ Paper
☐ Tempera paints
☐ Scissors

Sponge Painting

Create your own painting using the instructions on page 362 of the student book.

Biomimetics Challenge

Read the text on page 362 of your student book. What three things can we learn from biomimetics?

What did we learn?

1. How does a sponge eat?

2. How does a sponge reproduce?

3. Why is a sponge an animal and not a plant?

🚀 Taking it further

1. Why can a sponge kill a coral colony?

2. What uses are there for sponges?

3. Why are synthetic sponges more popular than real sponges?

| | God's Design: Life | The World of Animals | Day 167 | Unit 5 Lesson 31 | Name |

31 Worms
Creepy crawlers

🧪 Supply list
☐ Gummy worms
☐ Dirt
☐ Dried leaves
☐ Rocks
☐ Crushed chocolate cookies
☐ Instant chocolate pudding

🧪 Worm Diorama

Make a scene in a shoebox showing an earthworm's habitat. Include dirt, rocks, dried leaves, and any other items you might find where earthworms live. Use gummy worms for the earthworms. Take a picture of your diorama for your animal notebook.

🧪 Wormy Snack

Follow the instructions on page 364 of your student book to create a yummy snack!

🏅 Supplies for Challenge
☐ Paper
☐ Paint
☐ Drawing materials

🏅 Tube Worms Challenge

Read page 365 in your student book. Draw or paint a picture of this amazing ecosystem to include in your animal notebook. .

The World of Animals 297

What did we learn?

1. What kinds of worms are beneficial to man?

2. How are they beneficial?

3. What kinds of worms are harmful?

Taking it further

1. How can you avoid parasitic worms?

God's Design: Life | The World of Animals | Day 169 | Unit 6 Lesson 32 | Name

Kingdom Protista

Simple creatures?

🧪 Supply list

☐ Colored pencils, markers, or crayons
☐ Construction paper
☐ Yarn
☐ Shoe
☐ Scissors
☐ Glue
☐ Microscope (optional)
☐ Pond water (optional)
☐ Slides (optional)

🧪 Paramecium Model

Create a paramecium model by following the instructions on page 369 for this activity.

🧪 Observe Microscopic Creatures

Note: This activity is optional. If you have access to a few drops of water from a pond or stream and a microscope, examine the water and see what tiny creatures may live in it.

🏅 Sporozoans Challenge

Read the text on page 369 of the student book. Tell your teacher what you learned about sporozoans and why they are dangerous to people and animals.

🧠 What did we learn?

1. How are protists different from animals?

2. How are they the same?

🚀 Taking it further

1. Why is a euglena a puzzle to scientists?

2. Why are single-celled creatures not as simple as you might expect?

| God's Design: Life | The World of Animals | Day 171 | Unit 6 Lesson 33 | Name |

33 Kingdom Monera & Viruses
Good and bad germs

Supply list
☐ Hand soap and other anti-bacterial items in your house

Anti-bacterial Hunt
Follow the instructions on page 371 of the student book to get hints to find things labeled "anti-bacterial." How many did you find?

Antibiotic Resistance Challenge
Tell your teacher what you learned about antibiotic resistance. What does it mean that fossilized bacteria is very similar to today's bacteria?

What did we learn?

1. How are bacteria similar to plants and animals?

2. How are bacteria different from plant and animal cells?

3. How are viruses similar to plants and animals?

4. How are viruses different?

🚀 Taking it further

1. Answer the following questions to test if a virus is alive.

    Does it have cells?

    Can it reproduce?

    Is it growing?

    Does it move or respond to its environment?

    Does it need food and water?

    Does it have respiration?

    Is it alive?

2. How can use of antibiotics be bad?

| | God's Design: Life | The World of Animals | Day 174 | Unit 6 Lesson 34 | Name |

Animal Notebook

Putting the animals together

🧪 Final Project supply list

☐ Paper

☐ Art supplies

☐ Clip-art or other animal pictures

☐ Worksheets from previous lessons

☐ Photographs of projects from previous lessons

🧪 Animal Notebook

Follow the instructions and suggestions on page 376 of the student book to complete more sections of your animal notebook.

🎖 Invertebrate Collage

Demonstrate your knowledge of invertebrates by making a collage of pictures of various invertebrates to be included in your animal notebook. You can draw the pictures, cut them out of magazines, or print them from your computer. Try to include invertebrates from all six groups mentioned in the lesson. Use an animal encyclopedia or other source for ideas.

🧠 What did we learn?

1. What do all animals have in common?

2. What is the difference between vertebrates and invertebrates?

3. What sets protists apart from all the other animals?

🚀 Taking it further

1. What are some of the greatest or most interesting things you learned from your study of the world of animals?

2. What would you like to learn more about?

| | God's Design: Life | The World of Animals | Day 177 | Unit 6 Lesson 35 | Name |

 Conclusion

Reflecting on the world of animals

 Supply list

☐ Bible

☐ Paper and pencil

☐ Bible concordance or Bible encyclopedia

Read Job 38:39–40:5 and discuss all the wonders mentioned in this passage. Discuss Job's response to God's questions. How should we respond to God's creation?

Giving Thanks

Write a poem or prayer of thanksgiving to God for the amazing world of animals.

Optional:

Use a Bible concordance or Bible encyclopedia and see how many different animals you can find mentioned in the Bible.

Plant Quizzes and Final Exam
for Use with
The World of Plants
(*God's Design: Life* Series)

| 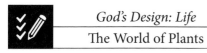 God's Design: Life
 The World of Plants | Quiz 1 | Scope: Lessons 1–4 | Total score: ____ of 100 | Name |

Introduction to Life Science

Note: While all students take the first portion of the quizzes, the questions on the next page are for students who have completed the challenges included in lessons 1-4. The challenge questions are also worth a total of 100 points.

Mark each statement as either True or False (7 points each).

1. _____ All living creatures have cells.

2. _____ Plants do not need oxygen.

3. _____ Growth and change can be signs of life.

4. _____ Nonliving things absorb nutrients.

5. _____ Plants cannot move, so they are not alive.

6. _____ A kingdom is a way to group things together by similar characteristics.

7. _____ Plants and protists are the two main kingdoms of living things.

8. _____ Plants and animals both have chlorophyll.

9. _____ Vacuoles store food inside of cells.

10. _____ The nucleus is the control center of a cell.

Short answer (15 points each):

11. Name three differences between plant cells and animal cells. _____

    _____

    _____

12. Describe how to tell if something is alive. _____

    _____

    _____

🏅 Challenge questions

Fill in the blanks using the terms below. Not all words are used (10 points each).

Biogenesis	Gymnosperm(s)	Mitosis	Telophase
Chemical evolution	Angiosperm(s)	Prophase	Cytokinesis
Spontaneous generation	Conifer	Metaphase	Monocot
	Ginkgo	Anaphase	Dicot

1. The law of _____ states that life always comes from life.

2. _____ says that life originally came from nonliving chemicals.

3. During _____ a cell divides into two identical cells.

4. A _____ is a type of plant that produces seeds in cones.

5. A _____ tree is sometimes called a living fossil.

6. _____ produce seeds that are enclosed in fruit.

7. The belief that life springs up from its environment is called _____.

8. During _____ the chromosomes in a cell line up in the middle.

9. During _____ the duplicate chromosomes are pulled apart.

10. _____ occurs when the cytoplasm in a cell is divided.

To calculate your grade if you have taken the challenge portion of this quiz:

_____ + _____ = _____ ÷ 2 = _____
(score on 1st part of quiz) (score on 2nd part of quiz)   (final score out of a possible 100%)

God's Design: Life — The World of Plants | Quiz 2 | Scope: Lessons 5–10 | Total score: ____ of 100 | Name

Flowering Plants & Seeds

Match the term with its definition (5 points each).

1. _____ Monocot
2. _____ Dicot
3. _____ Cotyledon
4. _____ Hilum
5. _____ Plumule
6. _____ Radicle
7. _____ Seed Coat
8. _____ Deciduous
9. _____ Evergreen
10. _____ Angiosperms

A. Seed with one cotyledon
B. Provides nourishment for the new plant
C. Protects the seed before germination
D. Seed with two cotyledons
E. Tree that loses its leaves in the winter
F. Where the seed was attached to the plant
G. Beginning of roots
H. Tree that does not lose its leaves in the winter
I. Beginning of stem and leaves
J. Trees that reproduce with flowers

Answer Yes or No. Would a seed germinate if placed in each of the following conditions? If no, explain what is missing (6 points each).

11. A seed planted in a garden in the spring time and watered every day _____

12. A seed planted in a desert _____

13. A seed planted in the dust on the moon _____

14. A seed in an envelope at the store _____

15. A seed in a moist paper towel in a sunny window _____

Short answer (20 points):

16. Name the four organs of a plant.

_____

_____

Note: Challenge questions are for students who have completed the challenges included in lessons 5-10. The challenge questions are also worth a total of 100 points.

🏅 Challenge questions

Mark each statement as either True or False (10 points each).

1. _____ Many plants can be used to make medicines.

2. _____ All grass is alike.

3. _____ Rye, wheat, and oats are all grasses.

4. _____ Many trees have distinctive crowns.

5. _____ A pine tree has a triangular growth habit.

6. _____ Pruning will not affect a tree's growth habit.

7. _____ External seed dormancy depends on temperature.

8. _____ Some commercial growers use sulfuric acid to scarify seeds.

9. _____ Stratification of seeds can occur in a refrigerator.

10. _____ Seeds with double dormancy can experience scarification and stratification in either order and still germinate.

To calculate your grade if you have taken the challenge portion of this quiz:

_____ + _____ = _____ ÷ 2 = _____
(score on 1st part of quiz)   (score on 2nd part of quiz)   (final score out of a possible 100%)

| God's Design: Life — The World of Plants | Quiz 3 | Scope: Lessons 11–15 | Total score: ____ of 100 | Name |

Roots & Stems

Choose the best answer for each statement or question (6 points each).

1. _____ Which is not a function of the roots of a plant?
   A. Absorb water   B. Produce seeds   C. Provide support   D. Store food

2. _____ Which is not considered a plant organ?
   A. Seeds   B. Stems   C. Flowers   D. Roots

3. _____ A plant with this type of roots is most likely to live where it is dry.
   A. Fibrous   B. Parasitic   C. Tap   D. Aerial

4. _____ A plant with this type of roots is most likely to live in a tree.
   A. Fibrous   B. Parasitic   C. Tap   D. Aerial

5. _____ You would be most successful planting plants with these roots on a steep hill.
   A. Fibrous   B. Parasitic   C. Tap   D. Aerial

6. _____ A plant with this kind of roots will be more successful in a very wet area.
   A. Fibrous   B. Prop   C. Tap   D. Aerial

7. _____ Which is not a function of the stem?
   A. Absorb water   B. Provide support   C. Transport water   D. Store food

8. _____ What causes water to move upward in a plant?
   A. Gravity   B. Sunshine   C. Transpiration   D. Phloem

9. _____ What is a new stem called?
   A. Stemling   B. Node   C. Bud   D. Shoot

10. _____ Where do leaves connect to the stem?
    A. Internode   B. Terminal bud   C. Node   D. Shoot

11. _____ Which kind of cells are not found inside a stem?
    A. Xylem   B. Epidermis   C. Phloem   D. Cambium

12. _____ Which cells carry water up a stem?
    A. Xylem   B. Epidermis   C. Phloem   D. Cambium

13. _____ Which cells carry food down a stem?
    A. Xylem	B. Epidermis	C. Phloem	D. Cambium

14. _____ Which cells protect a young stem?
    A. Xylem	B. Epidermis	C. Phloem	D. Bark

15. _____ Which cells protect mature stems?
    A. Xylem	B. Epidermis	C. Phloem	D. Bark

Short answer (10 points):

16. How do parasitic roots "steal" nutrients from nearby plants?

    _____

Note: Challenge questions are for students who have completed the challenges included in lessons 11-15. The challenge questions are also worth a total of 100 points.

Challenge questions

Choose the best answer for each question (20 points each).

1. _____ How does primary growth change the root?
   A. Becomes fatter	B. Becomes longer	C. Becomes shorter	D. Becomes lighter

2. _____ Where does primary growth occur in a root?
   A. Near root cap	B. In central core	C. In root hairs	D. In xylem

3. _____ How is osmosis different from diffusion?
   A. Slower	B. Requires a membrane	C. Faster	D. Occurs at night

4. _____ What role does capillarity play in plants?
   A. Makes food	B. Protects stems	C. Controls growth	D. Moves fluids up

5. _____ Which type of branching results in a wide, low tree or shrub?
   A. Excurrent	B. Terminal	C. Deliquescent	D. Scalar

To calculate your grade if you have taken the challenge portion of this quiz:

_____ + _____ = _____ ÷ 2 = _____
(score on 1st part of quiz) (score on 2nd part of quiz)   (final score out of a possible 100%)

| God's Design: Life The World of Plants | Quiz 4 | Scope: Lessons 16–20 | Total score: ____ of 100 | Name |

Leaves

1. What is the purpose of the stomata in leaves? **(10 points)** _____
   _____

2. For each leaf below, describe its vein arrangement (palmate or pinnate) and identify it if you can **(10 points each)**.

   A. _____   C. _____

   B. _____   D. _____

   A.

   B.

   C.

   D.

Short answer (10 points each):

3. Which leaf above appears to be a compound leaf? _____

4. What kind of leaf arrangement does plant D have? _____

5. How do leaves follow the sun? _____
   _____

6. What makes leaves green? _____
   _____

The World of Plants 315

7. Identify the "ingredients" (beginning materials) and the "products" (ending materials) of photosynthesis.

   Ingredients: _____

   Products: _____

Note: Challenge questions are for students who have completed the challenges included in lessons 16-20. The challenge questions are also worth a total of 100 points.

🏅 Challenge questions

Short answer:

1. For each leaf in question 2, describe the leaf margin **(15 points each)**.

   A. _____   C. _____

   B. _____   D. _____

2. Name two pigments that could be found in leaves **(5 points each)**.

   _____   _____

Short answer (10 points each):

3. What is the chemical formula for photosynthesis? _____

   _____

4. What is the purpose of a bract? _____

   _____

5. What is one purpose of a succulent leaf? _____

   _____

To calculate your grade if you have taken the challenge portion of this quiz:

_____ + _____ = _____ ÷ 2 = _____
(score on 1st part of quiz) (score on 2nd part of quiz)   (final score out of a possible 100%)

God's Design: Life
The World of Plants — Quiz 5 — Scope: Lessons 21–25 — Total score: ____ of 100 — Name

Flowers & Fruits

1. Identify the parts of the flower below **(10 points each)**.

   Ovary   Ovule   Petals   Pistil   Pollen   Sepal   Stamen

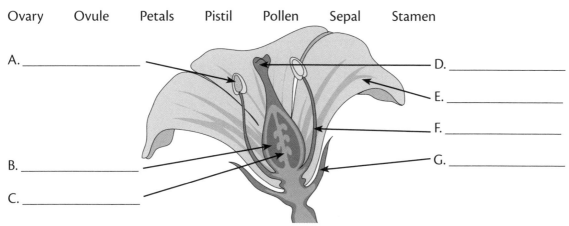

A. _____
B. _____
C. _____
D. _____
E. _____
F. _____
G. _____

Short answer (5 points each):

2. Which part of the flower is considered the male part? _____

3. Which part of the flower is considered the female part? _____

4. How is a simple fruit different from an aggregate fruit? _____

5. Describe the process of pollination. _____

6. List two ways that pollen can be transferred from one flower to another.

7. How long does it take for a biennial to complete its lifecycle? _____

Note: Challenge questions are for students who have completed the challenges included in lessons 21-25. The challenge questions are also worth a total of 100 points.

Challenge questions

Mark each statement as either True or False (10 points each).

1. _____ Composite flowers have only one flower per stalk.

2. _____ Ray flowers often look like petals.

3. _____ Disk flowers produce hundreds of seeds.

4. _____ Flowers do not need to attract pollinators.

5. _____ Some nectar guides can normally only be seen by insects.

6. _____ Some flowers smell bad to attract flies as pollinators.

7. _____ Succulent fruits are simple fruits.

8. _____ An olive is considered a fruit.

9. _____ Peanuts are nuts.

10. _____ Apples are berries from a biological definition.

To calculate your grade if you have taken the challenge portion of this quiz:

_____ + _____ = _____ ÷ 2 = _____
(score on 1st part of quiz) (score on 2nd part of quiz)   (final score out of a possible 100%)

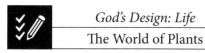

God's Design: Life — The World of Plants | Quiz 6 | Scope: Lessons 26–34 | Total score: ____ of 100 | Name

Unusual Plants

Match the term with its definition (5 points each).

1. ____ Plants that eat insects
2. ____ Plants that "steal" nutrients from other plants
3. ____ Plants that grow on other plants without harming them
4. ____ Response of plants to gravity (roots go down, stems go up)
5. ____ Tendency for roots to grow toward water
6. ____ Ability of leaves to turn toward sunlight
7. ____ Plant designed to store available water in dry conditions
8. ____ Plant reproduction using a part of the plant (not seeds)
9. ____ Runners from a strawberry plant
10. ____ Special stems that grow underground for reproduction

A. Geotropism
B. Parasites
C. Carnivorous
D. Passengers
E. Hydrotropism
F. Cactus
G. Phototropism
H. Stolons
I. Rhizomes
J. Vegetative reproduction

Short answer (10 points each):

11. What plant organ is missing in ferns? _____
12. How do both mosses and ferns reproduce? _____
13. What two plant organs are missing in mosses? _____
14. What substance do algae have in common with plants? _____
15. What is the name of the group that contains yeast and mushrooms? _____

Note: Challenge questions are for students who have completed the challenges included in lessons 26-34. The challenge questions are also worth a total of 100 points.

Challenge questions

Short answer (20 points each):

1. Give an example of positive tropism. _____
   _____

2. Give an example of negative tropism. _____
   _____

3. Explain how a cobra lily traps insects. _____
   _____

4. Explain how succulents are designed to survive dry periods. _____
   _____

5. Why is grafting a form of cloning? _____
   _____

To calculate your grade if you have taken the challenge portion of this quiz:

_____ + _____ = _____ ÷ 2 = _____
(score on 1st part of quiz) (score on 2nd part of quiz)   (final score out of a possible 100%)

| God's Design: Life / The World of Plants | Final Exam | Scope: Lessons 1–34 | Total score: ____ of 100 | Name |

Final Exam: World of Plants

Define each of the following terms (5 points each).

1. Geotropism: _____

2. Angiosperm: _____

3. Phototropism: _____

4. Photosynthesis: _____

5. Pollination: _____

6. Chlorophyll: _____

7. Ovule: _____

8. Pistil: _____

9. Stamen: _____

10. Xylem and phloem: _____

Choose the best answer for each question (2 points each).

11. ____ Which of the following is not an organ of flowering plants?
    A. Leaves   B. Roots   C. Chlorophyll   D. Stems

12. ____ Which kind or kinds of creatures get nourishment from grasses?
    A. Cows   B. Birds   C. Humans   D. All three (A, B, and C)

13. ____ Which is not a common use for the wood of a tree?
    A. Clothing   B. Fuel   C. Shelter   D. Calculating the tree's age

14. ____ Which tree is a deciduous tree?
    A. Pine   B. Maple   C. Spruce   D. Juniper

15. ____ Which organ is primarily used to absorb minerals from the ground?
    A. Flowers   B. Leaves   C. Roots   D. Stems

The World of Plants ✏️ 321

Fill in the blank with the correct term (2 points each).

16. Root growth primarily occurs at the _____.

17. The two types of root systems are _____ and _____.

18. The shape of most monocot plants' leaves is _____.

19. Ferns reproduce by _____ on their fronds.

20. Algae are similar to plants because they contain _____.

Mark each statement as either True or False (3 points each).

21. _____ Plants with red leaves have no chlorophyll.

22. _____ Trees can be identified by their leaves.

23. _____ Coniferous trees do not have leaves.

24. _____ The scent of a flower has no purpose.

25. _____ Ferns are not flowering plants.

26. _____ Algae is an important organism.

27. _____ Sepals might be confused with leaves.

28. _____ Pollination must take place for seeds to form.

29. _____ Mosses reproduce by spores and not seeds.

30. _____ Photosynthesis cannot take place without chlorophyll.

Note: Challenge questions are for students who have completed the challenges included in lessons 1-34. The challenge questions are also worth a total of 100 points.

🏅 Challenge questions

Match the term with its definition (5 points each).

1. _____ Law of biogenesis
2. _____ Meiosis
3. _____ Spontaneous generation
4. _____ Scarification
5. _____ Stratification
6. _____ Seed dormancy
7. _____ Primary growth
8. _____ Secondary growth
9. _____ Osmosis
10. _____ Toothed leaf margin
11. _____ Lobed leaf margin

A. Belief that life can come from nonlife
B. Seed will not germinate
C. Cell division for reproduction
D. Breaking of seed coat
E. Life can only come from life
F. Period of cold before germination
G. Diffusion through a membrane
H. Jagged leaf edge
I. Growth in width or circumference
J. Growth in length
K. Large indentations around the edge of leaf

Mark each statement as either True or False (5 points each).

12. _____ Ephemeral plants grow slowly.
13. _____ Composite flowers are really hundreds of flowers grouped together.
14. _____ Legumes have hard outer shells.
15. _____ Pomes have papery inner cores.
16. _____ Chemotropism aids in pollination.
17. _____ Filament algae is very common.
18. _____ Most commercial fruit trees are grown from seeds.
19. _____ Tendrils have negative tropism.
20. _____ Rootstock is important for grafting.

To calculate your grade if you have taken the challenge portion of this quiz:

_____ + _____ = _____ ÷ 2 = _____
(score on 1st part of quiz) (score on 2nd part of quiz)   (final score out of a possible 100%)

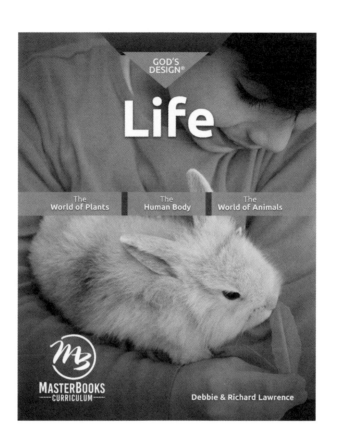

Anatomy Quizzes and Final Exam

for Use with

The Human Body

(*God's Design: Life* Series)

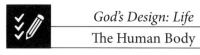 *God's Design: Life* — The Human Body | Quiz 1 | Scope: Lessons 1–3 | Total score: ____ of 100 | Name

Body Overview

Fill in the blanks with the correct term: (5 points each answer)

1. Many cells working together are called a _____.

2. _____ cells carry oxygen to the body.

3. _____ cells stretch and contract to allow for movement.

4. The _____ acts like the skin of a cell.

5. The _____ is the brain or control center of a cell.

6. Vacuoles are where cells store _____.

7. Mitochondria break down food to provide _____ for the cell.

8. Humans were created in _____ image.

9. The _____ system provides strength for the body.

10. _____ cells can be over a yard long.

11. Nutrients are provided to the body through the _____ system.

12. _____ protects the body from harmful substances outside the body.

13. _____ cells have a criss-cross shape.

14. Several tissues working together are called an _____.

15. A system is made up of _____, _____, and _____ all working together to perform a specific function.

Short answer (15 points):

16. Explain the function of red blood cells and white blood cells. _____

_____

_____

_____

Note: Challenge questions are for students who have completed the challenges included in lessons 1-3. The challenge questions are also worth a total of 100 points.

🏅 Challenge questions

Fill in the blanks with the correct term (20 points each).

1. The _____ system produces chemical messengers to control body functions.

2. Waste products are removed from the body by the _____.

3. A mother carries the unborn baby in her _____.

4. Bones are _____ tissue.

5. _____ tissue covers the inside of your stomach.

To calculate your grade if you have taken the challenge portion of this quiz:

_____ + _____ = _____ ÷ 2 = _____
(score on 1st part of quiz) (score on 2nd part of quiz)   (final score out of a possible 100%)

| God's Design: Life — The Human Body | Quiz 2 | Scope: Lessons 4–10 | Total score: ____ of 100 | Name |

Bones & Muscles

Place the letter for the correct bone type next to each description below (7 points each).

1. _____ Makes new blood cells
2. _____ Found in hands and feet
3. _____ Supports most of your weight
4. _____ Found in arms and legs
5. _____ Found in face
6. _____ Gives protection to organs
7. _____ Vertebrae
8. _____ Gives flexibility in hands
9. _____ Ribs and skull
10. _____ Shoulder blades

A. Short bones
B. Long bones
C. Flat bones
D. Irregular bones

Mark each statement as either True or False (6 points each).

11. _____ Muscles stretch and contract individually.
12. _____ Muscles can be damaged by tearing.
13. _____ Using muscles makes them stronger.
14. _____ Approximately 40% of your body weight is from muscles.
15. _____ What you eat does not affect your bones and muscles.

Note: Challenge questions are for students who have completed the challenges included in lessons 4-10. The challenge questions are also worth a total of 100 points.

Challenge Questions

Mark each statement as either True or False (10 points each).

1. _____ The arm and leg bones are part of the appendicular skeleton.

2. _____ The axial skeleton primarily provides form and strength.

3. _____ Blood-clotting cells are some of the first cells at the site of a fracture.

4. _____ Broken bones are usually weaker after they heal than before the break.

5. _____ It can take weeks for a bone to fully heal.

6. _____ Joints are designed to keep bones in place and to move freely.

7. _____ Individual muscle cells each contract to make a muscle contract.

8. _____ Muscles help blood move through the body.

9. _____ The heart is made up of skeletal muscle tissue.

10. _____ Only hands and feet have friction skin.

To calculate your grade if you have taken the challenge portion of this quiz:

_____ + _____ = _____ ÷ 2 = _____
(score on 1st part of quiz) (score on 2nd part of quiz)   (final score out of a possible 100%)

| God's Design: Life The Human Body | Quiz 3 | Scope: Lessons 11–18 | Total score: ____of 100 | Name |

Nerves & Senses

1. Name the five senses your brain uses to collect information about the outside world **(6 points per answer)**.

   _____

Match each part of the brain with its function (5 points each).

2. _____ Cerebellum   A. Controls growth

3. _____ Brain stem   B. Memory

4. _____ Spinal cord   C. Thought, language, and decisions

5. _____ Cerebrum   D. Movement

6. _____ Hippocampus   E. Connects the nerves in the body to the brain

7. _____ Pituitary gland   F. Controls life functions

Mark each statement as either True or False (5 points each).

8. _____ You can improve your intelligence by exercising your brain.

9. _____ Smells can bring back memories.

10. _____ You don't need to wear a bike helmet when you ride your bike.

11. _____ The left side of the brain controls the left side of the body.

12. _____ What you eat affects your brain.

13. _____ Reflexes are slower than normal signals to the brain.

14. _____ Your brain fills in for the blind spot in your eye.

15. _____ The louder the sound is the higher its amplitude.

The Human Body ✐ 331

Note: Challenge questions are for students who have completed the challenges included in lessons 11-18. The challenge questions are also worth a total of 100 points.

Challenge questions

Choose the best answer for each question below (20 points each).

1. _____ Which nervous system is the only one capable of higher level complex thought?
   A. Fish
   B. Human
   C. Monkey
   D. Dog

2. _____ Which neurons process and generate signals?
   A. Motor
   B. Sensory
   C. Association
   D. Conglomeration

3. _____ What is the function of myelin?
   A. Insulation
   B. Interpretation
   C. Coagulation
   D. Impulses

4. _____ What liquid is found in the middle of the eye?
   A. Retina
   B. Aqueous humor
   C. Cones
   D. Vitreous humor

5. _____ What part of the ear controls balance?
   A. Eustachian tube
   B. Cochlea
   C. Semi-circular canal
   D. Malleus

To calculate your grade if you have taken the challenge portion of this quiz:

_____ + _____ = _____ ÷ 2 = _____
(score on 1st part of quiz) (score on 2nd part of quiz)   (final score out of a possible 100%)

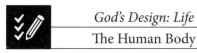 God's Design: Life — The Human Body | Quiz 4 | Scope: Lessons 19–23 | Total score: ____ of 100 | Name

Digestive System

1. Name the six major parts of the digestive system **(5 points each)**.

   _____   _____   _____

   _____   _____   _____

Match the type of tooth with its function (5 points each). Two types do the same function.

2. _____ Incisors   A. Tearing

3. _____ Canines   B. Cutting

4. _____ Bicuspids   C. Grinding

5. _____ Molars

Short answer (10 points per answer):

6. Why is it important to take good care of your teeth? _____

   _____

7. How do you take good care of your teeth? _____

8. Why is it important to eat healthy foods? _____

   _____

Short answer (2 points per answer):

9. Name the five different food groups you learned about. _____

   _____

10. Name the three forms of energy in food. _____

11. What two other important types of compounds do we get from our food?

    _____

Note: Challenge questions are for students who have completed the challenges included in lessons 19-23. The challenge questions are also worth a total of 100 points.

🏅 Challenge questions

Mark each statement as either True or False (10 points each).

1. _____ Enzymes play a crucial role in digestion.

2. _____ The gall bladder stores gastric juice.

3. _____ Pancreatic juice helps break down fats.

4. _____ Dentin gives the tooth its general size and shape.

5. _____ Enamel is one of the softest substances in the body.

6. _____ Orthodontics is the area of dentistry that corrects the alignment of teeth.

7. _____ A banana has fewer calories than a cup of french fries.

8. _____ Many diseases can be prevented by eating the right foods.

9. _____ Rickets can cause bleeding of the gums.

10. _____ Eating green, leafy vegetables can help prevent anemia.

To calculate your grade if you have taken the challenge portion of this quiz:

_____ + _____ = _____ ÷ 2 = _____

(score on 1st part of quiz) (score on 2nd part of quiz)   (final score out of a possible 100%)

| God's Design: Life The Human Body | Quiz 5 | Scope: Lessons 24–28 | Total score: ____ of 100 | Name |

Heart & Lungs

Fill in the blank with the part of blood that does each job below (10 points each).

1. _____ transport oxygen and carbon dioxide.

2. _____ surround and eliminate germs.

3. _____ carries the cells around the body.

4. _____ repair breaks in the blood vessels.

Match the term with its definition (6 points each).

5. _____ Diaphragm   A. Gas that exits the lungs

6. _____ Carbon dioxide   B. Tubes entering the lungs

7. _____ Oxygen   C. Muscle that expands the chest cavity

8. _____ Bronchi   D. Gas that enters the lungs

9. _____ Trachea   E. Tube in back of the throat through which air passes

Short answer (15 points each answer)

10. List two ways to keep your lungs healthy.

_____

_____

Note: Challenge questions are for students who have completed the challenges included in lessons 24-28. The challenge questions are also worth a total of 100 points.

Challenge questions

Short answer (20 points each):

1. What is the difference between systolic and diastolic blood pressure? _____
   _____

2. List two ways that a person might lower his/her blood pressure. _____
   _____

3. Describe how blood moves through the heart. _____
   _____
   _____

4. Which blood type or types can donate to someone with A positive blood? _____
   _____

5. Explain the difference between external and internal respiration. _____
   _____

To calculate your grade if you have taken the challenge portion of this quiz:

_____ + _____ = _____ ÷ 2 = _____
(score on 1st part of quiz)   (score on 2nd part of quiz)   (final score out of a possible 100%)

| God's Design: Life The Human Body | Quiz 6 | Scope: Lessons 29–33 | Total score: ____ of 100 | Name |

Skin & Immunity

Choose the best answer for each question (10 points each).

1. _____ Which of the following is not a part of your skin?
   A. Dermis   B. Esophagus   C. Epidermis   D. Subcutaneous tissue

2. _____ Which cells are found on the outermost part of the epidermis?
   A. Red blood cells B. Nerve cells   C. Muscle cells   D. Dead skin cells

3. _____ Where is friction skin found on your body?
   A. Hands   B. Feet   C. Both A and B   D. Neck

4. _____ Which of the following is not a function of the skin?
   A. Purifying blood B. Protection   C. Gripping   D. Temperature regulation

5. _____ Which people have the same fingerprints?
   A. Twins   B. Parent/children C. Siblings   D. No one

6. _____ Which organ provides a barrier against disease?
   A. Heart   B. Skin   C. Brain   D. Lungs

7. _____ Which of the following are produced by white blood cells?
   A. Antibodies   B. Red blood cells C. Spleen   D. Mucus

8. _____ Which of the following does not help filter germs from your body?
   A. Spleen   B. Digestive system C. Teeth   D. Lymph system

9. _____ How are physical traits passed on from parent to child?
   A. Genes   B. Blood transfusion C. Teaching   D. They are not

10. _____ Which of the following grows from a follicle?
    A. Tears   B. Nerves   C. Hair   D. Babies

Note: Challenge questions are for students who have completed the challenges included in lessons 29-33. The challenge questions are also worth a total of 100 points.

Challenge questions

Match the term with its definition (10 points each).

1. _____ Arrector pili muscles

2. _____ Albinism

3. _____ Double helix

4. _____ Base pair

5. _____ Thymine

6. _____ Cytosine

7. _____ Chromosome

8. _____ Deoxyribose

9. _____ Adenine

10. _____ Mutation

B. Tiny muscles in the skin attached to hairs

C. Shape of DNA

D. Sugar forming the sides of the DNA molecule

E. Forms a rung of the DNA molecule

F. Base used in DNA; must be paired with thymine

G. Base used in DNA; must be paired with adenine

H. A complete strand of DNA

I. A mistake in the genetic code

J. Base used in DNA; must be paired with guanine

A. Condition where the body does not produce melanin

To calculate your grade if you have taken the challenge portion of this quiz:

_____ + _____ = _____ ÷ 2 = _____
(score on 1st part of quiz) (score on 2nd part of quiz)   (final score out of a possible 100%)

| God's Design: Life The Human Body | Final Exam | Scope: Lessons 1–34 | Total score: ____ of 100 | Name |

Final Exam: Human Body

Choose the best answer (4 points each).

1. _____ Which of the following are the three main types of fingerprints?
   A. Whorls, arch, loop   B. Swirl, arch, hoop   C. Swoop, circle, line   D. Loop, hoop, curl

2. _____ What kind of skin is on the palms of your hands and bottoms of your feet?
   A. Slipping   B. Gripping   C. Friction   D. Smooth

3. _____ Which are not found in skin?
   A. Hair follicle   B. Alveoli   C. Heat receptors   D. Dermis

4. _____ Which is determined by genetics?
   A. Whom you marry   B. Eye color   C. Where you live   D. Your age

5. _____ Your hip is which type of joint?
   A. Saddle   B. Gliding   C. Hinge   D. Ball and socket

6. _____ What causes a muscle to contract?
   A. Signal from your brain   B. Pull from another muscle   C. Cooking   D. Being in a hot room

7. _____ Which is not a part of the nervous system?
   A. Brain   B. Nerves   C. Stomach   D. Spinal cord

8. _____ What is the main function of the hippocampus in the brain?
   A. Thinking   B. Short-term memory   C. Walking   D. Making your heart beat

9. _____ Which should you eat sparingly for good health?
   A. Breads and cereals   B. Fruit   C. Fats and sweets   D. Vegetables

Match the name of the system to its function (4 points each).

10. _____ Circulatory system   A. Breaks food down into nutrients

11. _____ Digestive system   B. Moves the bones in your body

12. _____ Skin   C. Gives support and protects vital organs

13. _____ Nervous system   D. Moves blood through the body

14. _____ Muscular system   E. Protects the body from germs

15. _____ Respiratory system   F. Senses things and sends messages to the brain

16. _____ Skeletal system   G. Takes air into the lungs and exhales it

Mark each statement as either True or False (4 points each).

17. _____ Long bones produce red blood cells.

18. _____ Your funny bone is a bone in your arm.

19. _____ It is important to drink plenty of water every day.

20. _____ You will get most of the vitamins you need if you eat a variety of foods.

21. _____ White blood cells stop a cut from bleeding too much.

22. _____ A human heart has four chambers.

23. _____ Your lungs are filled with branching tubes that end in alveoli.

24. _____ Your brain is the largest organ in your body.

25. _____ Skin contains pain and temperature receptors.

Note: Challenge questions are for students who have completed the challenges included in lessons 1-34. The challenge questions are also worth a total of 100 points.

🎖 Challenge questions

Mark each statement as either True or False (3 points each).

1. _____ The endocrine system produces hormones.

2. _____ The excretory system removes nutrients.

3. _____ The reproduction system was designed to create new life.

4. _____ Epithelial tissue moves muscles.

5. _____ Connective tissue holds the body together.

6. _____ The skeletal system is divided into axial and appendicular parts.

7. _____ Only humans have the ability to perform complex reasoning.

8. _____ Cartilage holds bones in place.

9. _____ Muscles only move bones.

10. _____ Reflexes are faster than other nervous system responses.

11. _____ Dendrites receive input.

12. _____ The Braille alphabet was developed to allow deaf people to communicate.

13. _____ Rods and cones allow you to eat ice cream.

14. _____ The cochlea is part of the middle ear.

15. _____ Enzymes speed up digestion.

Short answer (1 point per answer):

16. Name five bones in your body. _____

_____

17. Name five muscles in your body. _____

_____

18. Name four functions of the brain. _____

_____

19. List the four major blood types. _____

20. Name two diseases caused by poor nutrition. _____

_____

21. Name the three kinds of respiration. _____

_____

22. Name two purposes of melanin. _____

_____

Short answer (15 points each):

23. Based on what we learn from the Bible, God created all of creation as well as Adam and Eve without sickness or death. Why do we see diseases and death in the world today?

24. Scientists are amazed at how complex the human body is, and they learn new things about it every day. How does this complexity point to a Creator?

To calculate your grade if you have taken the challenge portion of this quiz:

_____ + _____ = _____ ÷ 2 = _____
(score on 1st part of quiz) (score on 2nd part of quiz)   (final score out of a possible 100%)

Animal Quizzes and Final Exam
for Use with
The World of Animals
(*God's Design: Life* Series)

| 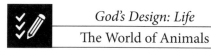 | *God's Design: Life* The World of Animals | Quiz 1 | Scope: Lessons 1–7 | Total score: ____ of 100 | Name |

Mammals

Short answer (10 points each):

1. What are the two main groups of animals? _____

2. What are the five major groups of vertebrates? _____
   _____

3. What are five common characteristics of mammals? _____
   _____

4. What makes a marsupial different from other mammals? _____

5. What makes a vertebrate unique? _____

Mark each statement as either True or False (5 points each).

6. _____ Animals can produce their own food.

7. _____ Dolphins are large fish.

8. _____ Marsupials give birth to tiny live babies.

9. _____ Baleen whales have large teeth.

10. _____ The elephant is the largest animal in the world.

11. _____ Monkeys have tails, but apes do not.

12. _____ Marsupials live primarily in Australia and Tasmania.

13. _____ Some marsupials are meat-eating animals.

14. _____ A lemur is a primate.

15. _____ Primates have eyes on the sides of their heads.

To calculate your grade if you have taken the challenge portion of this quiz:

_____ + _____ = _____ ÷ 2 = _____
(score on 1st part of quiz) (score on 2nd part of quiz)   (final score out of a possible 100%)

Note: Challenge questions are for students who have completed the challenges included in lessons 1-7. The challenge questions are also worth a total of 100 points.

Challenge questions

1. Draw a foot of a mammal to represent each of the following stances **(10 points each)**.

   Unguligrade   Digitigrade   Plantigrade

2. Match the parts of a ruminant's digestive system with its definition **(4 points each)**.

   _____ Rumen   A. First chamber of stomach

   _____ Abomasum   B. Second chamber of stomach

   _____ Cud   C. Food that is chewed a second time

   _____ Reticulum   D. Third chamber of stomach

   _____ Omasum   E. Fourth chamber of stomach

3. List three special design features that God gave whales **(5 points each answer)**.

   _____

   _____

   _____

4. List three special design features that God gave koalas **(5 points each answer)**.

   _____

   _____

   _____

5. Explain why an ape doing sign language does not necessarily support the evolution of man from apes **(20 points)**.

   _____

   _____

   _____

| God's Design: Life The World of Animals | Quiz 2 | Scope: Lessons 8–13 | Total score: ____ of 100 | Name |

Birds & Fish

Short answer (5 points each answer):

1. Look at each picture of birds' feet below. Below each picture write the term you think is most appropriate: **perching, water, runner, bird of prey**

   A._____ B._____ C._____ D._____

2. Look at each picture of birds' beaks below. Below each picture write what you think that bird is likely to eat: **seeds, nectar, other animals, water plants**

   A._____ B._____ C._____ D._____

Short answer (6 points each answer):

3. List three ways that birds were specially designed for flight.

   _____
   _____
   _____

4. List two special design features of a bird's digestive system.

   _____
   _____

5. Name three kinds of fins found on most fish.

   _____
   _____
   _____

6. Name two kinds of cartilaginous fish.

_____

_____

Note: Challenge questions are for students who have completed the challenges included in lessons 8-13. The challenge questions are also worth a total of 100 points.

🏅 Challenge questions

Mark each statement as either True or False (10 points each).

1. _____ Animals can adapt to changes in their surroundings.

2. _____ Birds evolved from reptiles.

3. _____ Birds and reptiles are both cold-blooded animals.

4. _____ Scales are very different from feathers.

5. _____ There can be great variety among species.

6. _____ Penguins only live in the southern hemisphere.

7. _____ Birds have a very efficient respiratory system.

8. _____ Birds have a bellows type of respiratory system.

9. _____ Fish scales fit together smoothly from head to tail.

10. _____ Fish have very small olfactory lobes compared to their brain size.

To calculate your grade if you have taken the challenge portion of this quiz:

_____ + _____ = _____ ÷ 2 = _____
(score on 1st part of quiz)   (score on 2nd part of quiz)   (final score out of a possible 100%)

| God's Design: Life The World of Animals | Quiz 3 | Scope: Lessons 14–19 | Total score: ____of 100 | Name |

Amphibians & Reptiles

1. What defines an animal as a vertebrate? **(10 points)** _____
_____

Place the letters of the characteristics that apply next to each animal group (18 points total per animal).

- A. Warm-blooded
- B. Cold-blooded
- C. Has hair or fur
- D. Has scales
- E. Has feathers
- F. Has no hair, feathers, or scales on its skin
- G. Gives birth to live young
- H. Lays eggs
- I. Experiences metamorphosis
- J. Has lungs
- K. Has gills
- L. Nurses its young
- M. Has wings
- N. Has fins

2. Mammals _____
3. Birds _____
4. Fish _____
5. Reptiles _____
6. Amphibians _____

Note: Challenge questions are for students who have completed the challenges included in lessons 14-19. The challenge questions are also worth a total of 100 points.

🏅 Challenge questions

Fill in the blank with the correct term below (10 points each).

Back	Plankton	Water	Carapace
Mouth	Legs	Pups	Apatosaurus
Air sacs	Frequency	Triceratops	
Sound	Plastron	Allosaurus	

1. Amphibians communicate primarily by _____.

2. Male frogs have inflatable _____ for communication.

3. Each frog species communicates on a different _____.

4. The Surinam toad presses eggs into the mother's _____.

5. The midwife toad carries eggs strapped to its _____.

6. The mouth-brooding frog carries its tadpoles in its _____.

7. A(n) _____ is a ceratopsian dinosaur.

8. A(n) _____ is a theropod dinosaur.

9. A(n) _____ is a sauropod dinosaur.

10. Marine iguanas are well adapted to life in and near the _____.

To calculate your grade if you have taken the challenge portion of this quiz:

_____ + _____ = _____ ÷ 2 = _____
(score on 1st part of quiz) (score on 2nd part of quiz)   (final score out of a possible 100%)

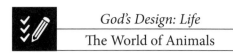

God's Design: Life
The World of Animals | Quiz 4 | Scope: Lessons 20–26 | Total score: ____of 100 | Name

Arthropods

Write *Yes* if the creature below is an arthropod. Write *No* if it is not (4 points each).

1. _____ Ant
2. _____ Tick
3. _____ Trout
4. _____ Spider
5. _____ Scorpion
6. _____ Crab
7. _____ Cricket
8. _____ Butterfly
9. _____ Clam
10. _____ Mouse
11. _____ Centipede
12. _____ Snail
13. _____ Roly-poly
14. _____ Crawdad
15. _____ Starfish
16. _____ Lizard

17. What are the four stages of complete metamorphosis for an insect?

_____   _____   _____   _____

Fill in the blanks with the appropriate numbers (2 points each answer).

18. An insect has _____ body parts, _____ legs, _____ antennae, and _____ wings.

19. A spider has _____ body parts, _____ legs, _____ antennae, and _____ wings.

20. A centipede has _____ pair(s) of legs per body segment and a millipede has _____ pair(s) of legs per body segment.

Note: Challenge questions are for students who have completed the challenges included in lessons 20-26. The challenge questions are also worth a total of 100 points.

Challenge questions
Short answer (10 points each):

1. Explain the purpose of an arthropod's exoskeleton. _____

_____

_____

2. What is the main ingredient in an exoskeleton? _____

3. Identify which segment (head, thorax, abdomen) is primarily responsible for each of the following functions in insects.

   Locomotion _____

   Internal functions _____

   Sensory input _____

4. What are two purposes of bioluminescence in fireflies?

   _____

   _____

5. Do male or female tarantulas live longer? _____

6. What do tarantulas usually eat? _____

7. List three types of symmetry commonly found among animals. _____

   _____

8. What does it mean if an animal is asymmetrical? _____

   _____

9. How does a millipede protect itself from predators? _____

   _____

10. Which is more dangerous, a centipede or a millipede? _____

To calculate your grade if you have taken the challenge portion of this quiz:

_____ + _____ = _____ ÷ 2 = _____
(score on 1st part of quiz)   (score on 2nd part of quiz)   (final score out of a possible 100%)

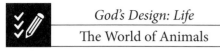

Other Invertebrates

Mark each statement as either True or False (5 points each).

1. _____ All mollusks have visible shells.

2. _____ A bivalve has two parts to its shell.

3. _____ You can identify a mollusk by the shape of its shell.

4. _____ The octopus is considered one of the least intelligent invertebrates.

5. _____ Cnidarians usually experience a polyp stage sometime in their lifecycle.

6. _____ Coral and algae have a symbiotic relationship.

7. _____ Echinoderms are usually very dark colors.

8. _____ Several invertebrates have the ability to regenerate.

9. _____ Echinoderms have smooth skin.

10. _____ A sponge is one of the simplest invertebrates.

11. _____ Sponges can reproduce by eggs.

12. _____ All worms are harmful to humans.

Short answer (8 points each):

13. What do jellyfish, coral, and sea anemones have in common? _____

_____

14. Name three groups of worms. _____

15. What part of the mollusk secretes its shell? _____

16. Which kind of mollusk has only one part to its shell? _____

17. What is an adult jellyfish called? _____

Note: Challenge questions are for students who have completed the challenges included in lessons 27-31. The challenge questions are also worth a total of 100 points.

Challenge questions

Short answer (10 points each):

1. Explain how cephalopods move. _____
   _____

2. How can a nautilus remain buoyant as its shell gets bigger and heavier? _____
   _____

3. What is a siphonophore? _____

4. Name a common siphonophore. _____

5. What is a madreporite in a starfish? _____

6. What technology is being improved by the study of the Venus Flower Basket sponge?
   _____

7. What is the name of the process that provides food for tubeworms?
   _____

Short answer (10 points each answer):

8. What is a symbiotic relationship? Describe two examples of cnidarians that have symbiotic relationships.
   _____

To calculate your grade if you have taken the challenge portion of this quiz:

_____ + _____ = _____ ÷ 2 = _____
(score on 1st part of quiz)   (score on 2nd part of quiz)   (final score out of a possible 100%)

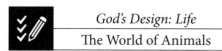

Simple Organisms

Match the parts of a cell to its function (10 points each).

1. _____ Nucleus
2. _____ Cell membrane
3. _____ Cytoplasm
4. _____ Mitochondria
5. _____ Vacuole

A. Cell's power plant
B. Liquid that facilitates transportation
C. Skin of the cell
D. Storage area
E. Brain of the cell

Match the single-celled organism with its description (10 points each).

6. _____ Flagellate
7. _____ Sarcodine
8. _____ Ciliate
9. _____ Bacteria
10. _____ Virus

A. Moves using pseudopods
B. Moves using tiny hairs over body
C. Moves using a whip-like tail.
D. Nonliving but contains strands of DNA
E. Cell with no defined nucleus

Note: Challenge questions are for students who have completed the challenges included in lessons 32-33. The challenge questions are also worth a total of 100 points.

🏅 Challenge questions

Mark each statement as either True or False (10 points each).

1. _____ Sporozoans have a very simple lifecycle.

2. _____ Sporozoans reproduce asexually and sexually.

3. _____ Sporozoans are parasites.

4. _____ Plasmodium is a dangerous protist.

5. _____ Antibiotic-resistant bacteria prove evolution.

6. _____ Survival of the fittest is the same as evolution.

7. _____ Fossilized bacteria are very similar to modern bacteria.

8. _____ Bacteria support biblical creation.

Short Answer (10 points each):

9. Many protists are parasitic and live in water. Name two of the serious diseases they can cause.

10. How do protists generally reproduce?

To calculate your grade if you have taken the challenge portion of this quiz:

_____ + _____ = _____ ÷ 2 = _____
(score on 1st part of quiz) (score on 2nd part of quiz)   (final score out of a possible 100%)

Final Exam: World of Animals

Match each animal group with its unique characteristic (3 points each).

1. _____ Mammals   A. Fins and gills

2. _____ Birds   B. Jointed feet or jointed legs

3. _____ Fish   C. Mantle that produces a shell

4. _____ Reptiles   D. Nurses its young

5. _____ Amphibians   E. Mostly single-celled

6. _____ Arthropods   F. Dry, scaly skin

7. _____ Mollusks   G. Feathers

8. _____ Echinoderms   H. Gills and lungs

9. _____ Cnidarians   I. Spiny skin

10. _____ Protists   J. Stinging tentacles

Define the following terms (6 points each).

11. Invertebrate: _____

12. Vertebrate: _____

13. Cold-blooded animal: _____

14. Warm-blooded animal: _____

15. Moneran: _____

Describe how a bird's feet are suited for each task listed below (4 points each).

16. Swimming in a lake: _____

17. Perching in a tree: _____

18. Hunting prey: _____

Short answer (13 points):

19. Describe how a bird is specially designed for flight. _____

_____

_____

Short answer (3 points each):

20. Name the three body parts of an insect.

_____

21. Name the two body parts of a spider.

_____

Mark each statement as either True or False (1 points each).

22. _____ Snakes have a special organ for sensing smell.

23. _____ Cold-blooded animals do not need to eat as often as warm-blooded animals.

24. _____ Turtles can safely be removed from their shells.

25. _____ Cartilaginous fish do not have any bones.

26. _____ Centipedes are arthropods.

27. _____ All crustaceans live in the water.

28. _____ Insects are the most common arthropod.

29. _____ Some creatures can live closely with jellyfish.

30. _____ The best way to kill a starfish is to cut it in half.

Note: Challenge questions are for students who have completed the challenges included in lessons 1-34. The challenge questions are also worth a total of 100 points.

Challenge questions

Match the term with its definition (4 points each).

1. _____ Unguligrade
2. _____ Digitigrade
3. _____ Plantigrade
4. _____ Rumen
5. _____ Abomasum
6. _____ Reticulum
7. _____ Omasum

A. First chamber of a cow's stomach
B. Walks on tips of toes
C. Second chamber of a cow's stomach
D. Walks on flats of toes
E. Walks on soles of feet
F. Third chamber of a cow's stomach
G. Fourth chamber of a cow's stomach

Mark each statement as either True or False (4 points each).

8. _____ Darwin's finches prove evolution.

9. _____ Birds have very efficient respiratory systems.

10. _____ Fish sense food by smelling the water.

11. _____ Frogs often confuse one species' call for another.

12. _____ The largest dinosaurs were the sauropods.

13. _____ Marine iguanas live only in the Galapagos Islands.

14. _____ The plastron is the top of a turtle's shell.

15. _____ A turtle's shell is made of the same material as fingernails.

Short answer (4 points each):

16. Exoskeletons are made from _____.

17. The legs of an insect are attached to the _____ section of its body.

18. Bioluminescence causes an animal to _____.

19. Tarantulas have barbed _____ that they kick at their enemies.

20. A _____ moves through the water using jet propulsion.

21. Biomimetics is the study of animals to apply designs to _____.

22. Tube worms live near _____.

23. Plasmodium causes the disease _____.

24. _____ are used to treat bacterial infections.

25. The stomach of a _____ is needed to complete the sexual reproduction of the plasmodium sporozoan.

To calculate your grade if you have taken the challenge portion of this quiz:

_____ + _____ = _____ ÷ 2 = _____
(score on 1st part of quiz) (score on 2nd part of quiz)   (final score out of a possible 100%)

Worksheet Answer Keys

for Use with

God's Design: Life **Series**

The World of Plants — Worksheet Answer Keys

Unit 1: Introduction to Life Science

1. Is It Alive?

What did we learn?

1. What are the six questions you should ask to determine if something is alive? **Does it eat? Does it "breathe"? Does it grow? Does it reproduce? Can it move? Does it have cells?**

2. Does the Bible refer to plants as living things? **How we classify plants in today's scientific world is different from how it was classified in the Bible. Also, there is a difference between people and animals and plants. While we consider both to be "living," plants are considered food and do not have the breath of life within them.**

Taking it further

1. Do scientists consider a piece of wood that has been cut off of a tree living? (Hint: Is it growing? Can it respond?) **No, it is not living anymore, although the tree it came from may still be living.**

2. Is paper alive? **No. It is made from wood but it is not alive.**

3. Is a seed alive? **This is a harder question. A seed has the potential for biological life, but it is not growing. You have to decide for yourself.**

2. What Is a Kingdom?

Clue Cards for Animal or Plant Game

1. Plants only—**Chlorophyll, photosynthesis, needs sun, cannot move around, needs carbon dioxide, created on the 3rd day of creation**

2. Animals only—**Moves around, cannot make food, carbon dioxide is a waste product, no chlorophyll**

3. Both—**Alive, cells, reproduces same kind, needs oxygen, designed by God, eaten by animals**

What did we learn?

1. What do plants and animals have in common? **God created them all, all are alive, all have cells, all reproduce their own kind, and all need oxygen**

2. What makes plants unique? **They have chlorophyll, perform photosynthesis, and cannot move freely.**

3. What makes animals unique? **They cannot produce their own food and can move freely.**

Taking it further

1. Are mushrooms plants? **No, they do not have chlorophyll or perform photosynthesis.**

2. Why do you think they are or are not? **Fungi have most of the characteristics of plants, but do not have chlorophyll and can live without sunlight. This is why scientists now group them in their own kingdom.**

3. Classification System

What did we learn?

1. What are the five kingdoms recognized today? **Plants, animals, fungi, protists, and monerans**

2. How do scientists determine how to classify a living thing? **They separate living things by common characteristics.**

3. What are the seven levels of the classification system? **Kingdom, phylum, class, order, family, genus, and species**

Taking it further

1. Why can pet dogs breed with wild wolves? **They are both the same kind of animal. Wolves, jackals, coyotes, wild dogs, and domestic dogs all came from the same ancestors. If any two animals can produce fertile offspring then they are most likely from the same animal kind. Wolves don't generally breed with domestic dogs because of their location and habits, but biologically they are the same kind of animal.**

2. How many of each animal did Noah take on the Ark? **Two of some animals and seven of other animals (see Genesis 7). Noah would only have taken two canines (dogs) on the Ark. Afterwards, the offspring of those two dogs resulted in the wide variety of dogs we see today.**

Challenge: Plant Classification

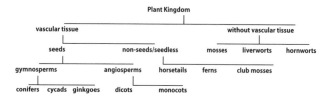

4. Plant & Animal Cells

What did we learn?

1. What parts or structures do all plant and animal cells have? **Cell membrane, nucleus, vacuoles, mitochondria, and cytoplasm**

2. What structures are unique to plants? **Cell wall and chloroplasts**

3. What distinguishes animal cells from plant cells? Possible answers: **Plant cells can perform photosynthesis and have cell walls but animal cells do not; their shape is different.**

Taking it further

1. A euglena is a single-celled living organism that can move around by itself. It eats other creatures, but it also has chlorophyll in its cell. Is it a plant, an animal, or something else? **Scientists do not agree on this and other unusual creatures. They usually put them in their own category, called protists.**

Unit 2: Flowering Plants & Seeds

5. Flowering Plants

What did we learn?

1. What are the four major parts of a plant? **Roots, stem, leaves and flowers**

2. What is the purpose for each part? **Roots hold the plant in place and suck up water and nutrients. Stems help the plant stand up and to move water and nutrients inside the plant. Leaves turn sunlight into food. Flowers produce seeds.**

Taking it further

1. What characteristics other than the flowers can be used to help identify a plant? **Leaves, fruit, and bark can all be used to identify a plant that is not in bloom.**

2. What similarities did you notice between the flowers you examined? **Answers will vary.**

3. What differences did you see? **Answers will vary.**

4. Can you use size to determine what a plant is? Why or why not? (Hint: Is a tiny seedling just as much an oak tree as the giant oak that is 100 years old?) **Size alone cannot tell you what a plant is.**

5. Why might you need to identify a plant? Possible answers: **To recognize poisonous plants such as poison ivy, to choose good plants for a garden or landscape, and to enjoy God's creation are just a few reasons**

6. Grasses

What did we learn?

1. Name four types of grass. **Turf, cereal, forage, and ornamental**

2. Describe the roots of a grass plant. **Fibrous root system with many small roots going out in several directions**

3. Why are grasses so important? **They are a major food source for animals and humans.**

Taking it further

1. Why can grass be cut over and over and still grow, while a tree that is cut down will die? **Recall that the leaves of the grass grow from the base of the plant. So cutting off the top of the leaves does not damage the growing center of the plant. However, trees grow at the ends of the stems and branches.**

2. Why is grass so hard to get rid of in a flower garden? **Consider the root structure; its fibrous design helps the plant spread and survive.**

3. What part of grass plants do humans eat? **They generally eat the seeds.**

4. What part of grass plants do most animals eat? **They usually eat the leaves and the seeds.**

5. Why can a cow eat certain grasses that you can't? **Cows have a very different digestive system that can break down the grass that humans can't digest.**

7. Trees

What Kind of Tree is This?

1. **Angiosperm, broad leaf, flowers, oak, maple, cherry should all have deciduous picture only. Gymnosperm, needles, cones, fir, pine, spruce, conifer should all have evergreen only. Seeds, bark, and growth rings should have both pictures. Note, a few evergreen trees have broad leaves so both pictures could be acceptable for this word. Also, a few trees that bear cones lose their leaves, so conifer could have both pictures.**

What did we learn?

1. What makes a plant a tree? **A single woody stem, grows to be tall, needs no support, and has bark**
2. How are deciduous and evergreen trees different? **Deciduous trees lose their leaves in winter and grow new ones in spring. Evergreen trees do not lose their leaves in the winter.**
3. How are angiosperms and gymnosperms different? **Angiosperms produce seeds with fruits and flowers. The seeds are located in the fruit. Gymnosperms produce seeds in cones.**

Taking it further

1. Do evergreen trees have growth rings? **Even though evergreens do not lose their leaves, they take a break from growing in the winter so they do have growth rings.**
2. How long do you think a tree lives? **Some trees only live a few years and some trees live to be hundreds, or even thousands, of years old. It depends on the variety of the tree and the growing conditions.**

8. Seeds

Seed Dormancy Challenge

1. condition where seeds do not germinate until certain conditions are met
2. the process in which the seed coat is broken or cracked
3. seeds that need stratification and scarification in order to germinate

What did we learn?

1. What conditions must be present for most seeds to sprout or germinate? **Water, oxygen, and warmth**
2. Why do seeds require these three conditions to begin growing? **God designed them that way so that the seeds will wait until it is likely the plants will survive before they germinate. If seeds germinated in the cold, the plant may not survive because the roots would freeze. If seeds germinated without oxygen, the plant would not be able to grow and it would die. If seeds germinated without water, the plants would soon wither. If seeds continued to germinate in unfavorable conditions, many plants could become extinct.**
3. Is soil necessary for seeds to germinate? **No, you germinated them in wet paper towels.**

Taking it further

1. If plants don't need soil to germinate, why do plants need soil to grow? **The seed has some stored energy that helps it get started. Once this energy is used up, the plant's roots must absorb nutrients from the soil, and the leaves will make food from the sun.**
2. Our seeds germinated in the dark. Can the plants continue to grow in the dark? **No, once the seed's energy is used up, the plant needs sunlight to make more food.**
3. How long can seeds remain dormant? **Some seeds have sprouted after 100 years or more of waiting for the right conditions.**

9. Monocots & Dicots

Seed dissection

1. How has the water affected the seed coat? **The water has softened the seed coat so that germination can begin.**
2. How does the seed coat of the corn differ from the seed coat of the beans? **The bean seed coat should be removed easily, and split into two parts. The corn seed coat cannot be removed easily, and will not divide.**

What did we learn?

1. What differences did you observe between the monocot and dicot seeds? **Monocots have one part and a tougher seed coat. Dicots have two parts and a softer seed coat.**
2. What parts of each seed were you able to identify? **You should have been able to identify the seed coat, hilum, plumule, radicle, cotyledons and the endosperm of the corn seed.**
3. What is the plumule? **Part that grows into the stem and leaves of the plant**
4. What is the radicle? **Part that grows into the roots of the plant**
5. What is the purpose of the cotyledon? **It provides energy for the new plant until the roots get big enough to begin supporting the plant, and sometimes becomes the first embryonic leaves of a seedling.**

Taking it further

1. Why did you need to soak the seeds before dissecting them? **Moisture is needed to soften the seed coat and begin the germination process. This is one of God's ways of preserving the seed until conditions are good for the plant to grow. If seeds sprouted when there was no water available, the young plants would soon wither and die.**

2. What differences do you think you might find in plants that grow from monocot and dicot seeds? **There are differences in the root structures, stems, leaves, and flowers. We will discuss some of these in later lessons.**

Challenge: Germination

- Explain to your instructor what the difference is between hypogeal and epigeal germination. **Hypogeal germination is where the cotyledons stay underground as the plant emerges (e.g., corn). Epigeal germination is where the cotyledons come up out of the ground and may appear to be leaves (e.g., beans).**

10. Seeds—Where Are They?

Seed Location Activity

Answers for 1-4 will vary, but the student should have chosen several fruits to be able to find and identify the types of seeds.

What did we learn?

1. What are three ways seeds can be moved or dispersed? **By the wind, by animals or by explosion**

2. Where are good places to look for seeds? **In fruit, pinecones, "dead" flowers, or pods**

Taking it further

1. How do people aid in the dispersal of seeds? Possible answers: **Farming, gardening, hiking (on our clothes), etc.**

2. What has man done to change or improve seeds or plants? Possible answers: **Man has used cross-pollination and genetic alteration of seeds to develop plants that are more resistant to disease or insects, that are higher in nutritional content, have larger blossoms or have different colors, as well as many other changes.**

3. If a seed is small, will the mature plant also be small? **Not usually; plants often grow large, even if they start from a small seed.**

4. Do the largest plants always have the largest seeds? **No**

5. Why do you think God created many large plants to have small seeds? **Small seeds are more easily dispersed.**

6. Can you name a plant that disperses its seeds by the whole plant blowing around? **Tumbleweed**

Challenge: Water Dispersal

- Why is seed dispersal important to plant survival? **Plants that are close together compete for room, nutrients, water, and sunlight. Seed dispersal allows plants to spread out. This gives the seeds a better chance to land in areas where there is more room, nutrients, water, and sunlight, thus allowing new plants a better chance of survival. The coconut can float because it is less dense than water, just like a ship.**

Unit 3: Roots & Stems

11. Roots

Root Observation Activity

1. Do carrots have a fibrous root or taproot system? **taproot**

2. Are carrot seeds more likely to be monocots or dicots? **dicot**

3. How are the roots of the beans and corn similar and how are they different? **Answers will vary based on the student's observations, but they should note both similarities (if any) and differences (if any).**

4. Were any corn or bean seeds planted "upside down"? Are the roots growing upside down? **Answers will vary based on how the student planted their seeds.**

Root Growth Challenge

1. List the three "zones" of cell division in a root. **zone of cell differentiation, zone of cell elongation, zone of cell division**

2. What is the name for the tube-like projections which absorb almost all the water and nutrients the plant needs? **root hairs**

3. Describe why roots are "specially designed" to move through soil. **Answers will vary, but should include a basic understanding of the following points: Roots are**

specially designed for moving through the soil. Not only do the cells add to the length of the root, but the root cap secretes a slimy substance that makes it easier for the root to move between the particles of soil. In addition, because the cells in the root tip are very active, they produce a relatively large amount of carbon dioxide. This combines with water in the soil to produce carbonic acid, which helps to loosen up the particles of soil around the root tip, making it even easier for the root to grow.

4. In which "zone" are root hairs? **zone of cell differentiation**

5. What are the two ways that a root grows? **primary and secondary root growth**

What did we learn?

1. What are the four organs of a plant? **Roots, stem, leaves, and flowers**

2. What are the jobs that the roots perform? **Anchoring the plant, absorbing water and nutrients, storing extra food**

3. How can you tell what kind of root system a plant has? **You can examine the roots. Also, look at the seeds to see how many cotyledons they have. Generally, monocots have fibrous roots and dicots have taproots. We will see in future lessons that monocots and dicots usually have different types of leaves, too. This can also give us a clue to what type of root structure a plant has.**

Taking it further

1. What kind of plants might you want to plant on a hillside? Why? **You might want to plant grass because the fibrous root system will spread out and hold the soil in place. Plants with taproots would not help stop the soil from washing away as well as plants with fibrous roots.**

2. Why are the roots of plants like carrots and beets good to eat? **The leaves produce more energy than the plant can use at one time so the extra energy is converted and stored in the roots. We can eat those roots and get that energy for our bodies.**

3. Why would a plant with a taproot be more likely to survive in an area with little rainfall than a plant with fibrous roots? **A taproot can generally grow much deeper than fibrous roots, so it would be more likely to reach underground water.**

12. Special Roots

Banyan Tree Challenge

1. What is another name for the banyan tree? **strangler fig**

2. How is it an epiphyte? **Its seeds are deposited on a host tree. It then grows on and often completely over the top of the host.**

3. Banyan trees are native to what country? **India**

4. Name two uses for the banyan tree. **Should include: meeting place, medical treatment/skin inflammation, polish for copper and bronze, gardening, rope, and paper**

What did we learn?

1. What are adventitious roots? **Roots that grow in unexpected places or in unexpected ways**

2. What are aerial roots? **Roots that grow in the air**

3. What are prop roots? **Roots that grow outward from the side of the stem, then downward to provide additional support**

Taking it further

1. Why do you think that some plants have specialized roots? **Not all environments are equally friendly to plant growth. God designed plants to grow in many areas that are hostile to most plants.**

2. Why do some plants need prop roots? **Plants that grow in very soft or often wet soil may not be anchored well enough with a single root system. They need prop roots to give additional support.**

13. Stems

Water Movements in Plants Challenge

osmosis, capillarity, and transpiration

What did we learn?

1. What are the main functions of a stem? **To support the plant, to carry nutrients and water between the roots and the leaves and flowers, to carry food from the leaves to the rest of the plant**

2. What do we call the stem of a tree? **A trunk or a branch**

Taking it further

1. If a tree branch is 3 feet above the ground on a certain day, how far up will the branch be 10 years later? **It will still be 3 feet above the ground. The tree will get taller but the growth is as the top of the tree and the ends of the branches, so the base of that particular branch remains in the same location.**

2. What are some stems that are good to eat? **Celery is a good stem. Potatoes are special stems called tubers. Onions are special stems called bulbs. These are a few examples of stems that are good to eat as vegetables.**

14. Stem Structure

What did we learn?

1. What are the major structures of a stem? **Shoot, terminal bud, lateral bud, node, and internode**

2. Where does new growth occur on a stem? **Most of the growth occurs at the terminal bud. New shoots, leaves, and flowers grow at lateral buds.**

3. What gives the plant its size and shape? **The collection and arrangement of stems**

Taking it further

1. What will happen to a plant if its terminal buds are removed? **It will stop growing. It may form new terminal buds on new shoots, but the existing stems will be unable to grow.**

2. How are stems different between trees and bushes? **Trees have one main stem or trunk with many smaller stems branching off. Bushes have many stems that all grow from the roots.**

3. In your experience, do flower stems have the same structures, including terminal buds, nodes, etc., as bush and tree stems? **Most plants have the same structures you learned about here. Sometimes the stems are very short and it may be difficult to identify all of the structures, but most of flowering plants have the same stem structures.**

15. Stem Growth

Looking at tree rings

1. Do you notice any rings that are significantly wider or narrower than the rest? **Yes, you can see differences in the tree rings.**

2. What can you guess about the growing condition when the rings are wider or narrower? **It is likely that the weather was wetter during the years with wider bands and drier during the years with narrower bands.**

3. Can you determine by the rings how old the tree is? **Everyone is likely to get a different number depending on where you start, but this tree was at least 70 years old.**

What did we learn?

1. What are epidermis cells? **The cells on the outside of a young stem.**

2. What is bark? **The epidermis cells that have been pushed outward, hardened, and died.**

3. Name three types of cells inside a stem. **Xylem, phloem, and cambium cells**

Taking it further

1. Can we tell a tree's age from the rings inside the trunk? Why or why not? **Different cells are produced during different parts of the growing season, and few cells are produced during the winter, so each year one set of colored bands is produced inside the trunk or stem of the tree. However, under certain conditions, multiple rings can form in a single year.**

2. If you wanted to make a very strong wooden spoon, which part of the tree would you use? **The center, where the heartwood is**

3. Why don't herbaceous plants have bark? **They die at the end of each growing season so there is no time for bark to develop.**

Challenge: Vascular Tissue

In lesson 13, you watched fluids moving up the stem of a stalk of celery. Do you remember how the xylem were arranged in the celery? They were arranged in a circular pattern. Would that indicate that celery is a monocot or a dicot? **Celery is a dicot.**

Unit 4: Leaves

16. Photosynthesis

What did we learn?

1. What are the "ingredients" needed for photosynthesis? **Water, carbon dioxide, chlorophyll, and sunlight**

2. What are the "products" of photosynthesis? **Food for the plant in the form of sugar and oxygen**

3. How did God specifically design plants to be a source of food? **Plants make their own food using the energy from the sun. They produce more than they need so the extra food energy is passed on to the animal or human that eats it.**

4. How does carbon dioxide enter a leaf? **Through holes in the leaf called stomata.**

Taking it further

1. On which day of creation did God create plants? **Day 3—Genesis 1:9–13**

2. On which day did He create the sun? **Day 4—Genesis 1:14–19**

3. In our experiment, we found that the plant that got less sunlight grew more slowly than the one that had full sunlight. Is this true for all plants? **No, many plants prefer the shade to full sun. God designed these plants to grow where other plants do not. Consider repeating this experiment with a shade-loving plant such as impatiens.**

17. Arrangement of Leaves

Special Leaves Challenge

What are the three special leaves that you learned about in this challenge? **Bracts, spines, succulent leaves**

What did we learn?

1. What are four common ways leaves can be arranged on a plant? **Opposite, alternate, whorled, and rosette**

2. Why do you think God created each of these different leaf arrangements? **Each of the leaf arrangements helps to ensure that one leaf does not block the light from reaching another leaf. The different arrangements are efficient for the size and shape of the leaves.**

3. Why is it important for sunlight to reach each leaf? **Leaves need sunlight for photosynthesis. This is what keeps the plant alive and growing.**

Taking it further

1. How does efficient leaf arrangement show God's provision or care for us? **Maximizing food production in plants provides more food for all animals, as well as for humans.**

2. What other feature, besides leaf arrangement, aids leaves in obtaining maximum exposure to sunlight? **Leaves turn to follow the sun. Cells away from the sun become longer than those on the sunny side, allowing the leaf to turn toward the sun as it moves through the sky.**

18. Leaves—Shape & Design

Observing Leaf Shapes and Vein Arrangements

1. How are the shapes and vein arrangements different? **Corn—long, thin, parallel; Bean—broad, palmate**

2. Which plant has broad leaves? **Bean**

3. Which has long narrow leaves? **Corn**

4. Which plant is a monocot? **Corn**

5. Which plant is a dicot? **Bean**

Leaf Shapes & Margins Challenge

1. margin
2. smooth
3. toothed
4. lobed
5. compound

What did we learn?

1. What general shape of leaves do monocots and dicots have? **Monocots usually have long narrow leaves with parallel veins. Dicots usually have wider flat leaves with either pinnate or palmate veins. Evergreens usually have needles or scales for leaves.**

2. How can we use leaves to help us identify plants? **Each plant has leaves with a unique shape and pattern. Some, such as the maple leaf, are very distinctive and easily recognizable. Others, grass for instance, can be more generic but can still be used to identify the species of plant.**

3. How do nutrients and food get into and out of the leaves? **Xylem brings nutrients into the leaf, and then after the leaf has performed photosynthesis, the phloem transports the sugar from the leaves to the rest of the plant.**

Taking it further

1. Describe how the arrangement of the veins is most efficient for each leaf shape. **Long, narrow leaves don't need veins that branch out, so parallel veins work well. Wider leaves need veins that reach out to the whole leaf, so palmate and pinnate veins are needed. For example, a maple leaf is nearly as wide as**

it is long, so a palmate arrangement of leaves is most efficient for transporting nutrients.

19. Changing Colors

What did we learn?

1. How do trees know when to change color? **The reduced amount of sunlight available each day signals the tree to begin preparing for winter.**
2. Why do trees drop their leaves? **As protection from the cold weather**
3. Why don't evergreen trees drop their leaves in the winter? **Their leaves are not easily damaged by the cold so the tree can survive the winter without losing them.**

Taking it further

1. Do trees and bushes with leaves that are purple in the summer still have chlorophyll? **Yes, but in smaller amounts compared to the red pigment.**
2. What factors, other than daylight, might affect when a tree's leaves start changing color? **Temperature can affect it to some extent. Also, the amount of water available has some effect, but even when the fall is unusually warm the leaves still begin to change about the same time each year because of the shorter period of daylight.**

20. Tree Identification

What did we learn?

1. What are some ways you can try to identify a plant? **By its leaves, flowers, fruit, bark, etc.**
2. What are the biggest differences between deciduous and coniferous trees? **Deciduous trees have flowers, broad leaves and lose all their leaves each fall. Coniferous trees have cones and needles, and do not lose their needles each fall.**

Taking it further

1. Why do we need to be able to identify trees and other plants? **To safely identify poisonous plants, to recognize ecosystems, for fun, to appreciate the diversity and wonder of God's creation and to make good landscaping choices**

Unit 5: Flowers & Fruits

21. Flowers

What did we learn?

1. What are the four parts of the flower and what is the purpose or job of each part? **The sepal protects the developing bud, the petals attract pollinators, the stamen produces pollen, and the pistil produces ovules that grow into seeds**

Taking it further

1. Why do you think God made so many different shapes and colors of flowers? **We enjoy the variety, and this shows us God's amazing creativity. Also, different animals are attracted by different colors and different scents. Hummingbirds are mainly attracted by red flowers, while other animals prefer different colors.**

22. Pollination

What did we learn?

1. What animals can pollinate a flower? **Bees, wasps, moths, hummingbirds, bats, beetles, and even some small rodents can all be pollinators.**
2. How can a flower be pollinated without an animal? **The wind or rain can move the pollen from the stamen to the pistil.**
3. Does pollen have to come from another flower? **Not always, but it generally comes from another flower on another plant**

Taking it further

1. Why do you suppose God designed most plants to need cross-pollination? **The genetic information is stored in the pollen and ovules. If plants were always self-pollinated, much genetic information would be lost. The seeds formed from cross-pollination combine the hereditary traits of both parents, and the resulting offspring generally are more varied and often healthier than would be the case with self-pollination.**

23. Flower Dissection

What did we learn?

1. How many ovules did you find? **Answers will vary.**
2. What did they look like? **Answers will vary.**

Taking it further

1. Why are the ovules in the flower green or white when most seeds are brown or black? **They are not fertilized and are not mature.**

2. If you planted the ovules, would they grow into a plant? **No, they are not pollinated and are not mature or ready to grow into a plant.**

Composite Flower Dissection

How do the flowers in the composite flower compare to the flower you previously dissected? How are they the same? How are they different? **Both kinds of flowers have stamens, pistils and petals. However, the composite flowers are much smaller, it is harder to see the reproductive parts, there are hundreds of little flowers connected to one stem.**

24. Fruits

What did we learn?

1. What is the main purpose of fruit? **To make sure the seeds are dispersed.**

2. What are the three main groups of fruit? **Simple, aggregate, and multiple**

3. Describe how each type of fruit forms. **Simple fruit forms one fruit from one flower with one pistil. Aggregate fruit forms one fruit from one flower with several pistils. Multiple fruit forms one piece from several flowers with each fruit fusing together into a whole.**

Taking it further

1. What is the fruit of a wheat plant? **The kernel of wheat that we make into flour**

2. Which category of fruit is most common? **Simple**

3. Why do biologists consider a green pepper to be a fruit? **They are the mature ovary of the plant. Any reproductive structure is a fruit. Other examples include beans, peas, tomatoes, and broccoli.**

Fruit Classification

1. _D_ Acorn
2. _E_ Pea
3. _C_ Pear
4. _A_ Avocado
5. _A_ Mango
6. _E_ Peanut
7. _E_ Lima bean
8. _F_ Wheat
9. _A_ Nectarine
10. _B_ Green pepper
11. _D_ Pecan
12. _C_ Crabapple
13. _B_ Grapefruit
14. _F_ Corn
15. _F_ Rice

25. Annuals, Biennials, & Perennials

What did we learn?

1. What is an annual plant? **One that completes its lifecycle in one growing season**

2. What is a biennial plant? **One that completes its lifecycle in two growing seasons**

3. What is a perennial plant? **One that grows year after year, producing flowers and seeds each season**

Taking it further

1. Why don't we often see the flowers of biennial plants? **We usually harvest them the first season.**

2. Why don't people grow new plants from the seeds produced by the annuals each year? **Often the conditions are not right for germination of the seeds produced the previous year. Also, many people clear their gardens of the dead flowers before the flowers have a chance to deposit their seeds. Finally, nurseries and greenhouses can begin growing plants inside much earlier than the plants would begin growing in the garden. This allows plants to be mature enough to be blooming by the beginning of spring. Plants that come up naturally in your garden would not bloom until much later in the summer, and people want to have blossoms in their gardens all spring and summer.**

Plant Word Search

```
P H O T O S Y N T H E S I S S
H R S E E D S K Q E R T A L P
O G T E E S A A N N U A L V A
S T E R B T M C H R Y S E E S
T B M P H O R O S Y N I A C H
C H L O R O P H Y L L M V O P
Z O P O N P U P U W R F E T E
L H D R M A B I P R U L S Y R
V I P O L L I N A T I O N L E
O R A O C M E N T L A W Q E N
P I N T E A N A I P T E C D N
A B C S E T N T B A I R O O I
L E V B W E I E C C A S T N A
M A N V C P A R E S N I A L L
A F R U I T L D E M F R I U R
```

The World of Plants 371

Unit 6: Unusual Plants

26. Meat-eating Plants

What did we learn?

1. What is a carnivorous plant? **One that eats animals, usually insects**

2. Why do some plants need to be carnivorous? **Some plants grow in areas where there are not enough nutrients in the soil. They trap insects to get the necessary nutrients to survive.**

3. How does a carnivorous plant eat an insect? **It traps the insect then secretes an acid that breaks down the animal's body so the nutrients can be absorbed.**

Taking it further

1. Where are you likely to find carnivorous plants? **In wet, marshy areas.**

2. How might a Venus flytrap tell the difference between an insect on its leaf and a raindrop? **The flytrap is designed with trigger bristles that have to be moved in order for the leaf to close. Two or more must be moved within a short period of time for the leaf to close. A raindrop might touch one but would probably not trigger two or more, but a moving insect would.**

27. Parasites & Passengers

What did we learn?

1. What is a parasitic plant? **One that gets its nourishment from another plant instead of making its own food.**

2. What is a passenger plant? **One that lives on the outside of another plant without harming it.**

3. How do passenger plants obtain water and minerals? **They absorb them from the air and from the surface of the host.**

Taking it further

1. Where is the most likely place to find passenger plants? **On trees—often in the rain forest, but also in other areas.**

2. Do passenger plants perform photosynthesis? **Yes, they still make their own food.**

3. Do parasitic plants perform photosynthesis? **Some do but most do not. They get their food from the host plant.**

28. Tropisms

What did we learn?

1. What is geotropism? **The ability of plant roots to always grow down and stems to grow up, a response to gravity**

2. What is hydrotropism? **The ability of plant roots to grow toward water, a response to water**

3. What is phototropism or heliotropism? **The ability of plant leaves to turn toward the sun or a light source, a response to light**

Taking it further

1. Why are tropisms sometimes called "survival techniques"? **They allow the plant to survive even if conditions change. They give the plant a better chance for survival even when water is scarce or something blocks the sun.**

2. Will a seed germinate if it is planted 5 feet (1.5 m) from the water? **No, seeds need water to soften the seed coat and germinate. Tropisms only help the plant after germination.**

3. Where are some places you would not want to plant water-seeking plants such as willows? **Near a swimming pool, septic tank, or water line.**

29. Survival Techniques

What did we learn?

1. How do some plants survive in hot, dry climates? **They quickly absorb the available water and store it in their expandable stems. They have needles instead of regular leaves so water does not evaporate quickly.**

2. How do some plants survive in cold, windy climates? **They have small leaves and short stems to withstand the wind. They grow low to the ground and in groups. They can reproduce very quickly.**

Taking it further

1. Why do alpine plants need protection from the sun? **High in the mountains there is less atmosphere, so the sun's rays are more intense.**

Challenge: Designed for Survival Worksheet

- List six things that plants need to survive. **Light, warmth, water, carbon dioxide, oxygen, minerals such as nitrogen and phosphorus, and a place to grow**

- List six things that can harm plants. **Wind, lack of water, hail, insects, diseases, over crowding, parasites, lack of light, and over watering**
- List 12 ways that plants have been designed to survive. **Broad leaf trees lose their leaves in winter; seeds do not germinate until conditions are favorable for growth (seed dormancy); plants have seed distribution techniques; seeds store energy for the growing shoot; special roots including prop roots, aerial roots, and pneumatophore roots; photosynthesis; leaf arrangement; leaf shape; flower shape; meat-eating plants' designs; special stems including tendrils, thorns, stolons, and runners; parasitic designs; passenger plants can use other plants to help them survive without harming them; all the various tropisms; ability to store water; needle-like leaves on cacti; design of alpine plants; bracts attract pollinators; scent, color, and pollen guides help attract pollinators**
- List four ways that people help plants to survive. **Water your grass, flowers, or other plants; add fertilizer to the soil; plant certain plants in a favorable location such as in the shade or sun depending on the plant; provide a trellis for climbing plants; pulling weeds to prevent competition; spray insecticide or fungicide; remove parasites; prune**

30. Reproduction Without Seeds

What did we learn?

1. What are some ways that plants can reproduce without growing from seeds? **Some plants reproduce by sending out runners, producing bulbs, or growing new plants from parts cut from the original plant.**

Taking it further

1. Why can a potato grow from a piece of potato instead of from a seed? **All of the growth information is located in the eyes of the potato, so a new plant can grow from this area.**
2. Will the new plant be just like the original plant? **Genetically, the new plant will be identical to the original plant. This is not the case with plants grown from seeds. Seeds contain genetic information from both the plant producing the pollen and the plant producing the ovule, but plants grown vegetatively only get genes from the original plant.**

31. Ferns

What did we learn?

1. How are ferns like other plants? **They have chlorophyll, stems, leaves, and roots.**
2. What are fern leaves called? **Fronds**
3. How are ferns different from other plants? **They do not have flowers or seeds.**
4. How do they reproduce? **They make spores on the back of their fronds that produce an egg and sperm that combine and grow into a tiny new plant.**

Taking it further

1. Why can't ferns reproduce with seeds? **They do not produce flowers, so they cannot make seeds.**

32. Mosses

What did we learn?

1. How do mosses differ from seed-bearing plants? **They have no flowers, seeds, or true roots.**
2. How do mosses differ from ferns? **They are smaller, have no true roots, and produce their spores on stalks instead of on their leaves.**
3. How do mosses produce food? **They have chlorophyll and perform photosynthesis just like other plants do.**

Taking it further

1. Are you likely to find moss in a desert? Why/why not? **No. There is not enough moisture in the desert for moss to grow well.**

33. Algae

What did we learn?

1. Why are algae such important organisms? **They produce large amounts of oxygen and they are the major source of food in many aquatic food chains.**
2. What gives algae its green color? **Chlorophyll.**

Taking it further

1. Why are some algae yellow, brown, blue, or red? **All algae have chlorophyll, but like many other plants, some have other pigments as well, which often cover up the green of the chlorophyll**

34. Fungi

Mold data sheet

What conditions were best for mold growth? **Warm, moist conditions should be best for mold growth. Also dark contributes to mold growth so sunlight may have affected your results.**

What did we learn?

1. Why are fungi not considered plants and given their own kingdom? **Fungi do not have chlorophyll and do not have roots, stems, and leaves as plants do.**

2. What are some good uses for fungi? **Fungi are used for food, to make bread rise, to make medicines, to give cheese its flavor, and to help in the recycling of dead plants and animals.**

Taking it further

1. What other conditions might affect mold growth other than those tested here? **Light/dark or the presence of chemicals like the preservatives found in foods**

2. How can you keep your bread from becoming moldy? **Keep it in a cool dry place.**

The Human Body — Worksheet Answer Keys

Unit 1: Body Overview

1. The Creation of Life

What did we learn?

1. On which day of creation did God make man? **On the sixth day**

2. In whose image did God create man? **In God's image**

3. According to Genesis 1:26, over what were man and woman to rule? **Fish, birds, livestock, all the earth, and all the creatures on the earth**

Taking it further

1. Since we are created in God's image, how should we treat our bodies? **We should take care of our bodies and keep them healthy.**

Challenge: Body Systems

Which system do you know the most about? Which system do you know the least about? Which system is the most interesting to you? **Answers will vary. The 11 systems of the human body include: skeletal, muscular, nervous, digestive, circulatory, respiratory, skin (integumentary), immune, endocrine, excretory, reproductive.**

2. Overview of the Human Body

What did we learn?

1. Name as many of the body's systems as you can and describe what each system does. **Answers will vary. See the list on page 142 and the Challenge section on page 143 of the student book.**

Taking it further

1. Which body systems are used when you walk across a room? **All of them. Your nervous, skeletal and muscular systems help you move. But also, your circulatory system provides oxygen and nutrients to your muscles. Your respiratory and digestive systems are what give your blood the oxygen and nutrients. And finally, your skin is needed to protect you as you walk.**

3. Cells, Tissues, & Organs

What did we learn?

1. What is the function of each of the following kinds of cells: skin cells, red blood cells, white blood cells, bone cells, nerve cells, and muscle cells? **Skin cells provide protection to seal out harmful substances and seal in moisture. Red blood cells carry oxygen to and carbon dioxide away from all the cells of the body. White blood cells eliminate invading germs and other harmful substances. Bone cells provide strength. Nerve cells relay messages. Muscle cells contract and expand to allow movement.**

Taking it further

1. How has God uniquely designed red blood cells to transport oxygen? **They are round and smooth so they easily flow through blood vessels. Also, their cell membranes allow oxygen and carbon dioxide to easily pass through.**

2. How are nerve cells specially designed to carry signals? **They have long tendrils or finger-like projections allowing a few cells to cover a large distance in the body, and thus signals can travel very quickly.**

3. How did God design skin cells to perform their special functions? **Their rectangular shape allows them to fit snuggly together, making an effective wall against germs and preventing moisture loss in your body.**

4. With all these cells working together, what do you think is the largest organ in the body? **You might guess the stomach, heart or brain, but your skin is actually the largest organ in your body.**

Challenge: Tissue Types

Skin: **Epithelial tissue**
Muscles: **Muscle tissue**
Tendons: **Connective tissue**
Lining of the mouth: **Epithelial tissue**
Brain: **Nerve tissue**
Inside of lungs: **Epithelial tissue**
Fat: **Connective tissue**
Bones: **Connective tissue**

Body Cells Worksheet

A. **Nerve cell**
B. **Muscle cell**
C. **White blood cell**
D. **Skin cell**
E. **Red blood cell**
F. **Bone cell**

Unit 2: Bones & Muscles

4. The Skeletal System

What did we learn?

1. What are three jobs that bones perform? **They provide strength, produce blood cells, and store calcium for future use.**

2. How are muscles connected to bones? **By cords called tendons**

3. What keeps bones from rubbing against each other at the joints? **A cushioning material called cartilage is between the bones.**

4. How many bones does an adult human have? **206**

5. What is the main mineral in bones? **Calcium**

Taking it further

1. What do you think is the largest bone in the body? **The femur or large leg bone is the longest and largest bone in the body.**

2. Why does this bone need to be so large? **The femur bones support most of the body's weight.**

3. What do you think are the smallest bones in the body? **The three bones in the inner ear—the malleus, incus, and stapes—are the smallest bones.**

5. Names of Bones

What did we learn?

1. Review the names of the bones by pointing to each bone as you name it.

2. Is your cranium above or below your mandible? **Your cranium, or skull, is above your mandible, which is your jaw bone. Of course, this is not true if you are standing on your head.**

3. What is moving if you wiggle your phalanges? **Your fingers and your toes.**

Taking it further

1. What happens if you cross your legs and gently hit just below your patella? **Your leg swings out with a reflex action. We will learn more about this when we study the nervous system.**

2. Why do we have Latin names for body parts? **This allows scientists to communicate about the body even if they do not speak the same language. This is why many scientific terms, not just body parts, are in Latin.**

Challenge: What's My Name? Worksheet

1. I hold your ribs together. **Sternum**
2. Your collar rests on me. **Clavicle**
3. Pat me on the back. **Scapula or vertebrae**
4. I support your weight when you stand. **Femur**
5. Listen, you might find us in a blacksmith shop. **Hammer, anvil, and stirrup.**
6. I'm in your leg, and that's no lie. **Fibula**
7. I rotate around your wrist. **Radius**
8. You wouldn't want to stub me. **Phalanges**
9. Man, I talk a lot. **Mandible**
10. When you cross your legs, I pop up. **Patella**

6. Types of Bones

What did we learn?

1. Which bones are designed mainly for protection of internal organs? **Flat bones such as your cranium and ribs**

2. Which type of bones helps determine what your face will look like? **Irregular bones in your face**

3. Which type of bones works closely with your circulatory system to replace old blood cells? **The long bones in your arms and legs**

Taking it further

1. Why are the long bones filled with marrow and not solid? **If they were solid, they would be too heavy to move easily. Also, the hollow design gives the bones more strength, since a cylinder is stronger than a solid rod. Finally, the marrow is where the red blood cells are produced.**

2. What is the advantage of having so many small bones in your hands? **This allows you to be flexible and move your fingers in lots of different ways so you can scratch your back, pick up a baby, or cook dinner.**

7. Joints

Scavenger Hunt

1. Door hinges: **Hinge**
2. Sliding doors: **Gliding**
3. Joints in pets: **Depends on the pet—dogs and cats have many of the same types of joints as humans including ball and socket, hinge, and others.**
4. LEGO® pieces: **May be able to find ball and socket, pivot, or saddle joints**

5. Nut crackers: **Hinge**
6. Pliers: **Pivot**

What did we learn?

1. What was the most common joint found around your house? **You probably found lots of hinges. This is the simplest kind of joint and is actually a modified lever and a very efficient way to move things.**

Taking it further

1. Which came first, the joints in the body or the joints in your house? **Obviously, the body was created first. Man realized the beauty and usefulness of God's designs and used them in many of his own inventions.**
2. Why do you need so many different kinds of joints in your body? **The different kinds of joints give our bodies lots of flexibility and strength, allowing us to move in so many different ways.**

8. The Muscular System

What did we learn?

1. How does a contracted muscle feel? **More rigid than a relaxed muscle**
2. How does a muscle get stretched? **After it relaxes, it is pulled by another muscle attached to the opposite side of a bone.**

Taking it further

1. How does a muscle know when to contract? **The brain sends a message to it telling it to contract.**
2. How does your face express emotion? **Your brain works together with your muscles to change the expression on your face. Muscles around your mouth and eyes move to show sadness, surprise, anger or happiness.**

9. Different Types of Muscles

What did we learn?

1. What are the two types of muscles? **Voluntary and involuntary**
2. How can we keep our muscles healthy? **By eating healthy foods, exercising, and stretching before you begin exercising**
3. How do your muscles learn? **Your brain automates functions that you do over and over again.**
4. What are some advantages of exercising? **Muscle strength, speed, endurance, and more energy**

Taking it further

1. Do you need to exercise your facial muscles? **Yes. Make funny faces at your mom.**

Challenge: Muscle Tissue

1. Diaphragm: **Striated**
2. Tongue: **Striated**
3. Esophagus: **Smooth**
4. Mother's womb: **Smooth**
5. Hand muscle: **Striated**
6. Heart: **Cardiac**

10. Hands & Feet

What did we learn?

1. Which is the most important finger? **The thumb**
2. Why is the thumb so important? **It is needed for grasping almost everything.**
3. What are some special features God gave to hands and feet? **They have flexibility because of multiple joints, sensitivity because of a vast array of nerves, special gripping skin, and protective nails.**

Taking it further

1. What activities or jobs require special use of the hands? **Musician, artist, and construction worker are just a few of the occupations requiring special use of the hands.**
2. What jobs require special use of the feet? **Most athletes, for example, soccer players and dancers**

Unit 3: Nerves & Senses

11. The Nervous System

What did we learn?

1. What are the three main parts of the nervous system? **Brain, spinal cord, and nerves**
2. In the response time test, what messages were sent to and from the brain? **The eye saw the object begin to fall and sent that message to the brain. The brain then sent a message to the hand and arm muscles telling them to contract.**

Taking it further

1. Name ways that information is collected by your body to be sent to the brain. **Eyes see, ears hear, your tongue tastes, your nose smells, and nerves in your skin send a variety of messages to your brain.**

Challenge: Unique Humans

Make your own list of things that humans can do that animals cannot do. What accounts for each of these abilities? **Worship God—humans have a soul and animals do not; feel true emotions—this is a complex thing involving the nervous system and the soul; humans train their bodies to do amazing physical feats, animals do not do this; people wear clothing; people are the only creatures that walk completely upright allowing them to freely use their hands—all of these attributes are related to our relationship with God.**

12. The Brain

What did we learn?

1. What are the three major parts of the brain? **Cerebrum, cerebellum, and brain stem**
2. Which part of the brain controls growth? **Pituitary gland**

Taking it further

1. Which part of the brain would be used for each of the following: running, dilating your eyes, learning your math facts? **The cerebellum controls muscles for running, the brain stem controls pupil dilation, and the cerebrum helps you learn new information like math facts.**
2. Is your brain the same thing as your mind? **No, your brain helps you think but your mind is more than just your thinking ability. Your mind is who God has created you to be. It includes your personality, and your soul and spirit, which allow you to have a relationship with God. The human soul is what truly sets man apart from the animals.**

Challenge: Brain Anatomy

1. Thought: **Frontal lobe, front part of cerebrum**
2. Smell: **Sensory area, central part of cerebrum**
3. Heart beat regulation: **Medulla oblongata, part of brain stem**
4. Memory: **Memory center on side of cerebrum as well as hippocampus**
5. Sight: **Visual cortex, at back of cerebrum**
6. Speech: **Speech center behind frontal lobe**
7. Muscle control: **Motor area in center of cerebrum as well as cerebellum**
8. Pupil dilation: **Midbrain, part of the brain stem**

13. Learning & Thinking

What did we learn?

1. Which part of the brain does each of the following: stores short-term memories, stores long-term memories, controls learning and thinking, controls the senses? **Hippocampus—short-term memory, cerebral cortex—long-term memory, cerebrum—learning/thinking, senses**
2. Which side of the brain controls the left side of the body? **The right side**
3. What is necessary for a healthy brain? **Good nutrition, sleep, and mental exercise**

Taking it further

1. List ways you can learn something. **By hearing, feeling, seeing, doing, smelling, and tasting. You actually will remember something best if you see it, hear it, say it back, and then associate it with something else.**
2. What is something you have trouble learning? **Come up with a new way to try to remember it. For example, if you have trouble remembering how to spell a word, write the letters you usually get wrong in a different color to help you see the right way to spell it.**

Challenge: Logic Puzzles

1. Four people must cross a river in a boat. Two people weigh 50 pounds each and the other two weigh 100 pounds each. They have a boat that can only hold a maximum of 100 pounds without sinking. Describe how all four people can cross the river in the boat. **The two 50 pound people go over together. One returns with the boat. One 100 pound person crosses over. The first 50 pound person returns with the boat. Again both 50 pounders cross over and one returns with the boat. The second 100 pound person crosses over. The second 50 pound person returns and picks up the first 50 pound person. This requires the boat to make 9 trips across the river.**
2. You are in a strange land and you need to find the nearest town. You know that one group of people living in the area always tells the truth and that the second group of people always tells a lie. You come to

a fork in the road and do not know which way to go. Standing at the fork are two people, one from each group, but you cannot tell which one is from which group. You can only ask one person one question. What question will you ask to be sure you take the correct road to reach the nearest town? **Ask either person, "If I ask the other person which road leads to the nearest town what would he say?" If this is the person that tells the truth, he will tell you the lie the other person would say. If this person is the one that tells a lie, he will tell you a lie even though the other person would have told the truth. Either way, the answer you get will be the wrong road to take so you can take the other road.**

14. Reflexes & Nerves

What did we learn?

1. How do reflex reactions differ from other nervous system messages? **They only go to the spinal cord not to the brain, so they are much faster.**

2. Why do we have reflexes? **They help us avoid dangerous situations.**

3. What are some different types of sensations detected by your nerves? **Texture, temperature, pain, vibration, and pressure**

Taking it further

1. What reflexes might you experience? **You duck if you sense something coming at you and close your eyes, or quickly pull back your hand when you touch something hot.**

2. How does the sense of touch differ from your fingertips to the back of your arm? **More nerves on the fingers allow you to detect more subtle differences.**

3. Why do you need a larger number of nerves on the bottoms of your feet? **To help you detect differences in the walking surface so you can keep your balance and not trip**

15. The Five Senses

What did we learn?

1. What are your five senses? **Sight, hearing, taste, smell, and touch**

2. Which of these senses usually gives us the most information? **Sight**

3. How does your brain compensate for the loss of one of your senses? **Your brain uses the other senses more to gather missing information.**

Taking it further

1. You have nerves all over your skin, so why don't you feel your clothes all day long? **Since the nerves detect the same feelings all day, your brain learns to ignore those messages so you don't notice your clothes. But if you really concentrate on it, you can feel your shirt rubbing against your arm.**

2. Your eyes see your nose all day long. Why don't you notice it all the time? **Since your eyes see your nose in your peripheral vision all the time, your brain learns to ignore that image, and it disappears. If you really try, you can see your nose. Cover one eye, and it will be more obvious.**

3. If you are in the hot sun for a while then you go inside, the room feels cold. Why? **Your brain is comparing the new temperature to the old temperature and decides it is cold. But after a few minutes, your brain becomes used to the new temperature and you don't feel cold anymore.**

Challenge: Braille System

Have student complete the activity on page 187 of the student textbook. **There are many more nerves in your fingertips than in the other parts of your hand, thus making it easier to feel different patterns of bumps.**

16. The Eye

What did we learn?

1. Name four important parts of the eye. **Lens, pupil, iris, retina, rods and cones, and optic nerve**

2. How does your brain compensate for different amounts of light in your surroundings? **By opening and closing the pupil—contracting and relaxing the iris**

3. How does your brain help you to focus on items that are near and items that are far away? **By adjusting the shape of the lens—by contracting muscles in the eye**

4. Why did God design our bodies with two eyes instead of just one? **Two eyes at different positions give depth perception. Also, if one eye is damaged the other can compensate.**

5. How does having two eyes help with a 3-dimensional image? **Each eye views an object from a different angle, allowing you to see more of the sides and giving you a better idea of the whole object.**

6. Since you have a blind spot, how can you see what is in that spot? **Your brain fills in with what is around the blind spot.**

Taking it further

1. Name some ways that the eye is protected from harm. **Eyelids and eyelashes keep out debris, tears wash away debris, the pupil contracts in bright light, the skull protects the eyes from impact**

2. Why do some people have to wear glasses or contact lenses? **The brain adjusts the lenses in the eyes to bring images into focus. Some people have lenses that are too flat or too round to be changed enough to make the images focus properly. Eyeglasses or contact lenses help compensate for these misshapen lenses in the eyes.**

3. Why can you fool your eyes or your brain into thinking you saw something you didn't actually see? **Your brain makes certain assumptions about what it expects to see based on what you normally see. If something is unusual you might be fooled, at least for a little while.**

Challenge: Liquid in Your Eyes

1. Lens: **Changes shape to focus on an image**
2. Pupil: **Dilates or closes to control amount of light entering eye**
3. Iris: **Controls size of pupil**
4. Cornea: **Front of the eye—protects lens**
5. Rods: **Detect light**
6. Cones: **Detect color**
7. Retina: **Contains rods and cones, detects image**
8. Optic nerve: **Nerve that transmits image to the brain**
9. Vitreous humor: **Keeps eyeball firm**
10. Aqueous humor: **Keeps front of eye firm, supplies nutrients to front of eye**

17. The Ear

What did we learn?

1. What characteristic of a sound wave determines how high or low a sound will be? **The frequency, or spacing, of the waves determines the pitch.**

2. What characteristic of a sound wave determines how loud or soft a sound will be? **The amplitude, or height, of the waves determines the volume.**

Taking it further

1. Why do two different instruments playing the same note at the same loudness sound different? **A note played by an instrument is not one single pitch. It actually has the same note played at various intervals called harmonics. For example, a C on the piano will sound very different than a C on a trumpet because of the different harmonics that are generated by the instrument.**

2. Name several ways to protect your hearing. **Ear plugs, ear phones, and turning down the volume**

3. How do you suppose deaf children learn to speak? **They watch other people to learn how to move their tongues, mouths and throats. They also can feel air movement from their mouths. Watch your mouth in a mirror as you say the sound of b and p. These sounds look the same. Then feel the air flow from your mouth as you say the same sounds. The p sound pushes out puffs of air but the b sound does not.**

4. How do you think a CD player or a telephone makes sounds? **Electronic devices such as telephones and CD players take electrical signals and send them through a device that vibrates, causing the air to move and thus making sounds from the electrical signals. This is the opposite of how your ears work, since your ear takes vibrations and turns them into electrical signals.**

Do You Hear What I Hear? Worksheet

_A_ Steaming tea kettle
_D_ Violin
_C_ Man singing
_B_ Bass drum
_C_ Falling snow
_A_ A jet engine
_D_ A TV show
_B_ A whisper

18. Taste & Smell

What did we learn?

1. What four flavors can your tongue detect? **Salty, sweet, sour, and bitter (and possibly umami)**

2. How does your tongue detect flavors? **Bits of food dissolved in saliva touch different taste buds in your mouth, and the taste buds generate electrical signals that go to the brain.**

3. How does your nose detect fragrances? **Scent particles enter your nasal cavity where smell-detecting nerves send signals to your brain.**

Taking it further

1. Can you still taste foods when you have a stuffy nose? **Depending on how bad your cold is, you may or may not be able to smell the foods. If your nose is really stuffy, the foods may taste bland and you may lose some of your appetite.**

2. Smells are used for more things than just enjoying food. List some other uses for your sense of smell. **Detecting dangerous odors such as smoke or gas, enjoying flowers, smelling the fresh air right after it rains.**

3. Oranges and grapefruits are both sweet and sour. Why do they taste different? **They have different proportions of sweet to sour. Also, they have different fragrances. This gives them each a unique flavor.**

4. Cocoa is very bitter. Why does chocolate candy taste so delicious? **The sweetness of the sugar combined with the bitterness makes it pleasant.**

Challenge: How We Taste and Smell

What is the input to each of your senses? **Sight—light/electromagnetic waves; sound—sound waves/mechanical waves/vibrating air; touch—direct stimulation of nerves in the skin; taste—chemicals in food; smell—odorants, which are also chemicals in food or other items**

Unit 4: Digestion

19. The Digestive System

What did we learn?

1. What are the main parts of the digestive system? **Teeth, tongue, esophagus, stomach, small intestine, and large intestine**

2. What role do your teeth play in digestion? **They grind and chop your food into small enough pieces to swallow.**

3. What role does your tongue play in digestion? **It helps move the food around in your mouth so you can chew it up, and it helps you swallow the food.**

4. Which is longer, your small intestine or your large intestine? **Your small intestine**

5. Which is wider, your small intestine or your large intestine? **Your large intestine**

Taking it further

1. Can you eat or drink while standing on your head? **Yes. You may think that gravity pulls the food into your stomach, but that is not the case. Involuntary muscles inside your esophagus push the food down,** so you can swallow even when you are upside down. You can test this by sipping some water through a straw while standing on your head. You may need some help to do this.

2. Why do some foods spend 30 minutes in the stomach while other foods spend 3 hours in the stomach? **Foods that are high in fat or protein take longer to break down than foods that are mostly starches.**

3. What makes you feel hungry? **When your stomach is empty, or nearly empty, it has nothing to move around, so it sends a message to your brain that you need more food. Only then do you feel hungry.**

4. Why did God design your body with a way to make you feel hungry? **Because humans are warm-blooded, they need to have a fairly constant supply of energy to help maintain their body temperature. Also, eating regularly helps us to have the energy necessary to do all the activities we like to do.**

Challenge: Digestive System Worksheet

1. _I_ Pancreas
2. _B_ Salivary glands
3. _A_ Tongue
4. _L_ Anus
5. _D_ Esophagus
6. _K_ Small intestine
7. _F_ Stomach
8. _E_ Liver
9. _G_ Gall bladder
10. _H_ Duodenum
11. _C_ Epiglottis
12. _J_ Large intestine

20. Teeth

What did we learn?

1. What is the job of each kind of tooth? **Incisors are for biting and cutting, canines are for tearing, and bicuspids and molars are for grinding.**

2. Why do people have baby teeth and why do they fall out? **You need small teeth when your mouth is small. Teeth can't grow like the rest of your body so the small teeth fall out, making room for larger teeth as your mouth gets larger.**

Taking it further

1. Why do we need to take care of our teeth? **If you don't take care of them, your teeth can get holes in them (cavities), break, or even fall out, making it hard to eat and causing you pain.**

21. Dental Health

What did we learn?

1. What are three things you can do to have healthy teeth? **Brush regularly, floss regularly, eat healthy foods, and visit the dentist regularly**

2. How does brushing your teeth help keep them healthy? **Brushing removes plaque and reduces the likelihood of getting cavities.**

3. List some foods that are good for your teeth. **Fruits, vegetables, and milk**

4. List some foods that are bad for your teeth. **Hard candy, sugary gum, and sipping sweet drinks for a long time**

Taking it further

1. Since your baby teeth are going to fall out anyway, why do you still need to brush them and take care of them? **You need to develop good habits even if you only have baby teeth. You will have a combination of baby and permanent teeth for several years. Also, even if your teeth are not damaged, bacteria can cause damage to the gums if you never clean your teeth.**

22. Nutrition

What did we learn?

1. What are the five food groups listed in this lesson? **Breads/grains, fruits, vegetables, dairy products, meat/beans**

2. What types of foods should you eat only a small amount of each day? **Junk foods such as cookies, candy, chips, and soda**

3. Why is variety in your diet important? **Different foods have different nutrients so you should eat a variety in order to get all the different things your body needs.**

Taking it further

1. Can a vegetarian eat a balanced diet? Hint: What other foods contain proteins found in meat? **Vegetarians can carefully combine plant products, such as dried beans, legumes, and nuts to obtain the same proteins as in meats.**

2. Is it necessary to eat dessert to have a healthy diet? **No, but it depends on what you consider dessert. Pie is not a great choice, but a banana would make a good dessert.**

Challenge: Nutrition Worksheet

Answers will vary. It is likely that the French fries and hamburger will have more calories, fat, and salt than the other foods. Which food is best for you depends on how you define what is good to eat. Fruits and yogurt are generally lower in calories, fat, and salt, so are probably better for you, but spaghetti with tomato sauce can be very healthy as well.

23. Vitamins & Minerals

What did we learn?

1. What are the three main forms of energy found in food? **Carbohydrates, proteins, and fats.**

2. How can we be sure to get enough vitamins and minerals in our diet? **The best way is to eat a variety of foods.**

3. Why is water so important to our diet? **Our bodies have a lot of water in them and use water for many different functions. Also, we lose water when we sweat and when we breathe, so we need to replace that water every day.**

Taking it further

1. Can you drink soda instead of water? **Soda and other drinks have water in them and the body can use that water. However, too much soda or other sweet drinks can give us more calories than we need. Also, soda may have a lot of salt in it, which can be unhealthy if you have too much. Finally, drinks containing sugar or salt can actually make you thirstier!**

2. Are frozen dinners just as healthy as fresh food? **Frozen dinners generally have a lot more fat, salt and calories than their fresh-made counterparts. They may be quicker and easier but they are not necessarily healthier.**

3. Is restaurant food as healthy as home-cooked food? **It depends on the restaurant. Some restaurants have salad bars and offer lots of choices for healthy foods. However, much restaurant food, especially fast food, is high in fat and calories, and limited in vegetables and fruits.**

Unit 5: Heart & Lungs

24. The Circulatory System

What did we learn?

1. What are the three main parts of the circulatory system? **Heart, blood, and blood vessels**
2. What are two functions of blood? **Supplying food and oxygen and removing waste products**
3. What are three types of blood vessels? **Veins, arteries, and capillaries**
4. Which blood vessels carry blood away from the heart? **Arteries**
5. Which blood vessels carry blood toward the heart? **Veins**
6. What happens to the blood in the capillaries? **Oxygen leaves the red blood cells and enters the surrounding tissue. Then, carbon dioxide enters the red blood cells to be taken to the lungs.**

Taking it further

1. How is the circulatory system like a highway? **Red blood cells are like delivery trucks because they carry oxygen, nutrients, and carbon dioxide to various parts of the body. Valves are like traffic signals because they keep blood flowing in the right direction. The oxygen, nutrients, and carbon dioxide are the cargo that gets transported. We will learn in later lessons how white blood cells are like policemen and platelets are like a construction crew.**
2. Why is exercise important for your circulatory system? **It strengthens your muscles, including your heart.**
3. List two other systems that depend on the circulatory system to function properly. **The digestive and respiratory systems both depend heavily on the circulatory system. Actually, every part of your body depends on the circulatory system.**
4. Why does your pulse increase when you exercise? **You need more oxygen when you exercise so your heart beats faster to get your blood moving faster, thus giving your body more oxygen.**

25. The Heart

What did we learn?

1. What are the four chambers of the heart? **Right and left atrium and right and left ventricle**
2. How many times does a blood cell pass through the heart on each trip around the body? **Two times—once before going to the lungs and once when returning from the lungs**

Taking it further

1. What are some things you can do to help your heart stay healthy? **Eat healthy foods, exercise, and don't smoke**
2. Is your heart shaped like a valentine? **No, it is shaped more like a grapefruit about the size of your fist.**
3. Does Jesus live in your physical heart? **No. The phrase "ask Jesus into your heart" is not found in the Bible. When a person is saved through repentance and faith, the Spirit of Christ (the Holy Spirit) dwells in his or her heart (Galatians 4:6; Ephesians 3:14–17). However, this "heart" does not refer to the physical organ, but to the "inner man." Jesus cannot live in someone's heart as He is seated at the right hand of the throne of God (Ephesians 1:20; Hebrews 8:1).**

The Heart Worksheet

A. **To arms and head**
B. **From arms and head**
C. **To lungs**
D. **From lungs**
E. **From legs and lower body**
F. **To legs and lower body**

26. Blood

What did we learn?

1. What are the four parts of blood and the function of each part? **Plasma transports the blood cells; red blood cells carry oxygen, carbon dioxide, and nutrients; white blood cells fight germs and other foreign substances; and platelets close wounds.**
2. What does your body do to help protect itself if you get cut? **Platelets swarm to the cut and make a patch.**
3. Do you have more red or white blood cells? **Many times more red blood cells than white blood cells.**

Taking it further

1. What are some of the dangers of a serious cut? **You can lose too much blood or get an infection.**

Challenge: Blood Types 1 Worksheet

Blood Donors

A	B	AB	O
A	B	AB	A
AB	AB		B
			AB
			O

Blood Recipients

A	B	AB	O
A	B	A	O
O	O	B	
		AB	
		O	

Challenge: Blood Types 2 Worksheet

Blood Donors

A+	A-	B+	B-	AB+	AB-	O+	O-
A+	A+	B+	B+	AB+	AB+	A+	A+
AB+	A-	AB+	B-		AB-	B+	A-
	AB+		AB+			AB+	B+
	AB-		AB-			O+	B-
							AB+
							AB-
							O+
							O-

Blood Recipients

A+	A-	B+	B-	AB+	AB-	O+	O-
A+	A-	B+	B-	A+	A-	O+	O-
A-	O-	B-	O-	A-	B-	O-	
O+		O+		B+	AB-		
O-		O-		B-	O-		
				AB+			
				AB-			
				O+			
				O-			

Which blood type is the universal donor and which is the universal recipient? **O is the universal donor, AB is the universal recipient. If you include Rh factor, the universal donor is O negative and the universal recipient is AB positive.**

27. The Respiratory System

What did we learn?

1. Describe the breathing process. **Your diaphragm contracts, expanding the chest cavity. Air fills the lungs and gases are exchanged. Then your diaphragm relaxes and the air exits the lungs.**

2. How do the circulatory and respiratory systems work together? **The circulatory system moves the blood around the body. In the lungs, the respiratory system exchanges gases with the blood.**

3. Where inside the lungs does the exchange of gases occur? **In the capillaries surrounding the alveoli**

4. What are the major parts of the respiratory system? **Nose, nasal passage, throat or pharynx, trachea, bronchi, lungs, alveoli, and diaphragm**

Taking it further

1. How do you suppose your body keeps food from going into your lungs and air from going into your stomach when both enter your body in the back of your throat? **You have a flap of tissue that covers the opening to the trachea when you swallow but opens up when you breathe.**

2. How does your respiratory system respond when you exercise? **Your brain senses that you need more oxygen when you are exercising, so it instructs your body to take more breaths. So you breathe faster.**

3. How does your respiratory system respond when you are sleeping? **Your brain senses that you need less oxygen when you are asleep than when you are awake, so it instructs your body to take slower, deeper breaths than when you are awake.**

The Respiratory System Worksheet

A. **Nasal cavity**
B. **Throat**
C. **Lung**
D. **Nose**
E. **Alveoli**
F. **Bronchial tube**
G. **Trachea**
H. **Diaphragm**

28. The Lungs

What did we learn?

1. How does your body keep harmful particles from entering your lungs? **Hairs in the nasal passage filter particles, tissues in the throat trap and kill bacteria and bronchial tubes are lined with mucous that traps particles**

2. How are your lungs similar to a balloon? **They both get bigger when filled with air and smaller when the air goes out**

3. How are your lungs different from a balloon? **A balloon is an empty sac; lungs have thousands of tiny sacs at the end of a series of branching pipes**

Taking it further

1. What can you do to keep your lungs healthy? **Exercise, try to avoid being around people who are sick if possible, and don't smoke**

2. If you breathe in oxygen and breathe out carbon dioxide, how can you help someone who is not breathing by breathing into his or her lungs when you do CPR? **Air going into the lungs is about 21% oxygen and 0.04% carbon dioxide. Air exiting the lungs is about 16% oxygen and 4% carbon dioxide. So when you breathe into someone else's lungs there is still quite a bit of oxygen in the air you exhale into their lungs.**

Unit 6: Skin & Immunity

29. The Skin

Skin Pigment Challenge

1. What is the most important role of melanin? **The most important role is protecting your skin from the harmful ultraviolet radiation from the sun.**

2. What color is the pigment carotene? **orange**

3. If your body cannot produce melanin, you have what condition? **albinism**

What did we learn?

1. What are the purposes of skin? **To protect you from the outside world, to keep your insides in and moist, and to hold the nerves that collect information for your brain**

2. How does skin help you stay healthy? **Primarily, it helps keep out germs**

3. How does skin allow you to move without getting stretched out? **It has elastin; an elastic substance that stretches then returns to its normal shape**

Taking it further

1. Other than skin, in what other ways does your body keep out germs? **Your nose and mouth have mucus and saliva that trap germs. Earwax helps to keep out germs. Hairs in your nose and ears help to keep out foreign objects. Scabs and white blood cells also keep out or get rid of germs.**

2. What skin problems might you experience in a dry climate? **Skin can get too dry and crack open, creating painful sores.**

3. What skin problems might you experience in a very moist or humid climate? **Your skin might secrete too much oil, causing oily skin. This can lead to acne.**

4. What are the dangers of a serious burn? **A serious burn takes away your skin and allows germs in. This can lead to infection and serious illness.**

30. Cross-section of Skin

What did we learn?

1. What are the three layers of skin? **Epidermis, dermis, and subcutaneous tissue**

2. What is the purpose of the sebaceous gland? **To secrete oil to keep the skin soft and flexible**

3. In which layer are most receptors located? **The dermis**

Taking it further

1. Explain what happens to your skin when you pick up a pin. **Nerves detect the size and shape of the pin and send a message to your brain. Your brain sends a message to your fingers to pick it up. As you pick it up your skin stretches and bends to the shape of the pin.**

2. How does your skin help regulate your body temperature? **Your skin secretes sweat when you are hot, which cools your body as it evaporates. Also, blood vessels in your skin get wider when you are hot to allow more heat to reach the surface of your skin. When you are cold, the blood vessels get smaller to help keep heat in. Fat cells in the subcutaneous tissue also help to insulate your body from temperature changes around you.**

Skin Word Search

F	**E**	**P**	**I**	**D**	**E**	**R**	**M**	**I**	**S**	S	I	D	O	I
S	K	N	E	D	F	T	**F**	**E**	Q	E	R	D	S	J
A	N	I	O	P	R	T	**A**	**L**	**P**	**D**	**F**	A	L	G
U	**E**	**S**	**U**	**B**	**C**	**U**	**T**	**A**	**N**	**E**	**O**	**U**	**S**	D
S	T	**K**	F	R	M	S	E	**S**	**W**	**R**	**L**	K	J	E
F	O	I	L	L	M	N	B	**T**	I	M	**L**	O	D	R
P	O	**N**	H	H	**B**	E	R	I	N	**I**	I	S	E	V
S	E	**K**	**E**	**R**	**A**	**T**	**I**	**N**	**G**	**S**	**C**	A	S	A
V	**G**	C	X	F	**R**	O	U	U	**E**	S	L	F	D	R
Y	**L**	R	I	T	**R**	T	**N**	**E**	**R**	**V**	**E**	**S**	K	R
N	**A**	T	D	G	I	B	U	G	**M**	F	K	H	I	L
E	**N**	S	M	I	E	L	K	F	**S**	**W**	**E**	**A**	**T**	D
R	**D**	I	W	**P**	**R**	**O**	**T**	**E**	**C**	**T**	**I**	**O**	**N**	G
A	**S**	A	R	E	W	**I**	O	P	Y	M	I	T	T	H
G	F	J	K	Y	T	**L**	L	T	O	I	Y	T	E	F

31. Fingerprints

What did we learn?

1. Where can friction skin be found on your body? **The palms of your hands and the bottoms of your feet.**

2. When are fingerprints formed? **About 3–4 months before birth**

3. What are the three major groups of fingerprints? **Arch, loop, and whorl**

4. Can you identify identical twins by their fingerprints? **Yes, even identical twins have unique fingerprints.**

Taking it further

1. What are some circumstances where fingerprints are used? **They are used to identify suspects in a crime, dead bodies, and lost children.**

2. Why do prints only occur on the hands and feet? **These are the only areas where friction skin is needed.**

3. Do children's fingerprints match their parents' prints? **No, each person has unique prints that do not match either parent.**

32. The Immune System

What did we learn?

1. What are the major parts of the immune system? **Skin, lymph, and circulatory systems**

2. What are the two major types of "germs" that make us sick? **Viruses and bacteria**

3. How do tears and mucus help fight germs? **Tears break down bacteria; mucus traps them.**

Taking it further

1. Why are a fever and a mosquito bite both indications that your immune system is working? **Your brain raises your body temperature in an effort to kill the germs. Swelling around a mosquito bite shows that antibodies are attacking the invading substance.**

33. Genetics

What did we learn?

1. What are genes? **Bits of information contained in each cell that control your physical development and many other things about you**

2. Why do children generally look like their parents? **Because half of the information that determines what a child looks like comes from the biological mother and half comes from the biological father.**

Taking it further

1. If parents look very different from each other what will their children look like? **It depends on the dominant genes. Generally, siblings look a lot like each other even if their parents look very different, but they have the potential to look very different. A couple in England had a very interesting family. The mother had very dark skin and the father had very light skin. They had twins and one had very dark skin and one had very light skin. The combinations of our genes make each of us a totally unique individual.**

2. In the past, evolutionists claimed that man evolved independently in different parts of the world and this is where the races came from. If this were true, how likely would it be that the different races could have children together? **This is very unlikely. God created all humans from one man and woman. More recent genetic testing indicates that all humans share a common ancestor just a few thousand years ago, which supports the teachings of the Bible.**

34. Body Poster

What did we learn?

1. Name the eight body systems you have learned about. **Skeletal, muscular, nervous, digestive, circulatory, respiratory, immune, and the skin (integumentary system)**

2. How do some of the different systems work together? **The circulatory system works with the respiratory system to bring oxygen to the body. The circulatory system works with the digestive system to bring energy to the whole body. The skin works with the nervous system to detect what is going on around us. The muscular system moves the skeleton. They all work together.**

Taking it further

1. What other systems can you think of that are in your body but were not discussed in this book? **Reproductive and excretory systems are a couple that were not covered, as well as the endocrine system. These systems are briefly discussed in the Challenge sections. You can learn more from an anatomy book.**

2. How do you see evidence of God the Creator in the design of the human body? **Answers will vary.**

The World of Animals — Worksheet Answer Keys

Unit 1: Mammals

1. The World of Animals

What did we learn?

1. What are the two major divisions of animals? **Vertebrates and invertebrates**
2. What are two similarities among all animals? **They move and they must eat plants or other animals.**

Taking it further

1. When did God create the different animal kinds? **On Day Five of creation God created fish and birds; on Day Six He created land animals. See Genesis 1.**
2. How is man different from animals? **Humans have a conscience so they can tell right from wrong; animals act on instinct. People have a spirit so they can have a relationship with God; animals do not. People were made in God's image; animals were not. Despite our physical similarities, people are spiritually very different from animals. God gave man dominion over the animals. See Genesis 1:28.**

2. Vertebrates

What did we learn?

1. What are the two major divisions of the animal kingdom? **Vertebrates and invertebrates**
2. What characteristics define an animal as a vertebrate? **Vertebrates have a spinal cord ending in a brain protected by a backbone. They also have internal skeletons.**
3. What are the five groups of vertebrates? **Mammals, birds, amphibians, reptiles, and fish**

Taking it further

1. Think about pictures you have seen of dinosaur skeletons. Do you think dinosaurs were vertebrates or invertebrates? Why do you think that? **Dinosaurs were vertebrates. This is shown by the fact that dinosaurs have internal skeletons and these skeletons contain vertebrae along the backs of the animals.**

3. Mammals

What did we learn?

1. What five characteristics are common to all mammals? **They are warm-blooded, breathe with lungs, give birth to live young, nurse their young, and have hair or fur.**
2. Why do mammals have hair? **To keep them warm, to aid in the sense of touch, and for some it provides camouflage.**
3. Why is a platypus considered a mammal even though it lays eggs? **It nurses its young.**

Taking it further

1. Name some ways that mammals regulate their body temperature. **Mammals cool down by sweating or panting. They heat up by eating, exercising, or covering their bodies to keep warm.**
2. What are some animals that have hair that helps them hide from their enemies? **Tigers and zebras have stripes that make them hard to see. Lions are the color of their surroundings.**

Challenge: Mammal Feet

1. Deer: **Unguligrade**
2. Rabbit: **Plantigrade**
3. Giraffe: **Unguligrade**
4. Wolf: **Digitigrade**
5. Skunk: **Plantigrade**
6. Elephant: **Unguligrade**
7. Opossum: **Plantigrade**
8. Chimpanzee: **Plantigrade**
9. Fox: **Digitigrade**

4. Mammals: Large & Small

What did we learn?

1. What is the largest land mammal? **Elephant**
2. What is the tallest land mammal? **Giraffe**
3. What do bears eat? **Nearly anything, including small animals and fish, but they prefer plants and berries.**

Taking it further

1. What do you think is the most fascinating mammal? Why do you think that? **Answers will vary.**

5. Monkeys & Apes

What did we learn?

1. What are two common characteristics of all primates? **They have ten fingers and ten toes, and eyes on the fronts of their heads giving them binocular vision.**

2. What are the three groups of primates? **Monkeys, apes, and prosimians**

3. What is one difference between apes and monkeys? **Apes do not have tails, monkeys do. Also, an ape's arms are longer than its legs, but this is not true for monkeys.**

4. Where do New World Monkeys live? **In the Western Hemisphere**

5. Where do Old World Monkeys live? **In the Eastern Hemisphere—Africa and Asia**

6. What is a prehensile tail? **A tail that can grasp**

Taking it further

1. If a monkey lives in South America is it likely to have a prehensile tail? **Yes, because New World Monkeys have prehensile tails and Old World Monkeys do not.**

2. Are you more likely to find a monkey or an ape in a tree in the rainforest? **You are more likely to find a monkey in a tree. Many apes do not spend a lot of time in trees, whereas most monkeys live the majority of their lives in trees.**

3. Why do most prosimians have very large eyes? **The majority of prosimians are nocturnal, that is, they sleep during the day and are awake at night. Large eyes allow these animals to see better at night.**

Mammals Word Search

```
D G E U T A F R O P K L H Y U
B E N D R M L J I O S C X F A
A Z E B R A Z C V B C A B S R
M N I D S M F V G F T T J A E
S R E I H M U X A A I M U P U
C O L S W A R M B L O O D E D
C A M E L R H O P T W U P E A
K E Y M O Y U S P Z E S R R W
F R L U G G D R W R G E K H U
M L U L E L I V E B I R T H S
O U E N A A T G L A R M D O B
C N H M O N K E Y V A I A N P
F G Y B U D L E W C F A N T L
H S A L E R U P Q Z F L U H E
R A J C J K W H A L E E S O P
```

6. Aquatic Mammals

What did we learn?

1. Why are dolphins and whales considered mammals and not fish? **They are warm-blooded, give birth to live young, nurse their babies and breathe air with lungs.**

2. What is the main difference between the tails of fish and the tails of aquatic mammals? **Fish tails move from side to side and mammal tails, or flukes, move up and down.**

3. What is another name for a manatee? **Sea cow**

4. Why are manatees sometimes called this? **They move slowly and graze on sea grass and other sea plants just like a cow grazing in a field.**

Taking it further

1. How has God specially designed aquatic mammals for breathing air? **First, He gave them blowholes or nostrils on the tops of their heads so it is easy to breathe while still being in the water. Second, He designed them to be able to stay submerged for several minutes or even an hour at a time so they do not have to stay near the surface. God also gave them flukes to help them resurface quickly.**

2. What do you think might be one of the first things a mother whale or dolphin must teach a newborn baby? **One of the first things the mother will do is push the baby toward the surface of the water so it can get its first breath.**

7. Marsupials

What did we learn?

1. What is a marsupial? **An animal that gives birth to very tiny underdeveloped young. The young then spend the next several months developing in the mother's pouch.**

2. Name at least three marsupials. **Some of the more common marsupials include kangaroos, koalas, opossums, numbats, and Tasmanian devils.**

3. How has God designed the kangaroo for jumping? **A kangaroo has large powerful hind legs, large hind feet and long stretchy tendons that help conserve energy when hopping.**

Taking it further

1. About half of a kangaroo's body weight is from muscle. This is nearly twice as much as in most animals its size. How might this fact contribute to its ability to hop? **Large muscles are needed to provide the strength to hop long distances, so a kangaroo has very large leg muscles.**

2. How do you think a joey kangaroo keeps from falling out of its mother's pouch when she hops? **The nipple swells when the joey first attaches so it cannot slip off. Also, the pouch has muscles that can contract like a drawstring to keep the pouch closed.**

Koala Fun Facts

1. We often hear a koala referred to as a koala bear, but it is not a bear. List three ways that a koala is different from a bear. **Koalas have pouches and give birth to very immature/undeveloped young; koalas only live in Australia; koalas live in trees and bears do not; koalas do not hibernate or sleep through the winter.**

2. What is unusual about the koala's pouch? **It opens toward the back.**

3. What does a baby koala eat? **Its mother's milk**

4. What does an adult koala eat? **Eucalyptus leaves**

5. What special design features make koalas able to eat this kind of food? **They have special grinding teeth, they have bacteria in their stomachs that help them digest the leaves.**

6. How long does a baby koala spend in its mother's pouch? **6 months**

7. How does the mother koala carry her youngster after it leaves the pouch? **On her back**

8. What is special about the skin on the koala's feet that helps it to climb trees? **It is rough—friction skin—that grips even smooth trees.**

9. What is special about the koala's hands that help it survive? **It has thumbs that grasp trees and leaves.**

10. How much does a koala sleep? **Up to 20 hours a day**

11. About how much does an adult koala weigh? **19–33 pounds**

12. What is the average life span of a koala? **10–15 years**

Unit 2: Birds & Fish

8. Birds

What did we learn?

1. How do birds differ from mammals? **Birds have feathers and wings, lay eggs, and usually can fly.**

2. How are birds the same as mammals? **They are both warm-blooded and breathe with lungs.**

Taking it further

1. How can you identify one bird from another? **By their size, shape, color of feathers, beak and feet shape, calls, and songs**

2. What birds can you identify near your home? **Use a field guide to help you.**

3. Why might you see different birds near your home in the summer than in the winter? **Many birds migrate to live in a warmer area in the winter and a cooler area in the summer, so different birds may be in your area at different times of the year.**

Challenge: Birds vs. Reptiles

List some characteristics that may vary among a species due to natural selection. Look through an animal encyclopedia to see examples of these characteristics. Notice that not one of these various characteristics has resulted in a new species. **Color, size and shape of beaks, size and shape of legs, size of body, coloring of feathers**

9. Flight

What did we learn?

1. What are some ways birds are designed for flight? **They have strong breast muscles, rigid backbones, hollow bones, efficient respiratory systems, feathers and wings.**

2. What are the three kinds of bird feathers? **Down, contour, and flight feathers**

3. How does a bird repair a feather that is pulled open? **By preening—running the feather through its beak to re-hook the barbs**

4. How does a bird's tail work like a rudder? **It is moved from side to side to help steer.**

Taking it further

1. Why can't man fly by strapping wings to his arms? **Man is not designed for flight. He does not have the strong breast muscles and stiff backbone needed. Humans are also too heavy to lift themselves with their arms.**

2. How do you think birds use their feathers to stay warm? **Birds can fluff up their feathers and trap air under and between them. The heat from their bodies warms the trapped air, creating a barrier between their bodies and the cold air around them.**

3. How is an airplane wing like a bird's wing? **They both have the same airfoil shape that allows the air flowing over the wing to create lift. Also, airplane wings are designed with the ability to change shape for different conditions, just like birds' wings.**

God Designed Birds to Fly Worksheet

Pictures should be similar to those in the student manual.

10. The Bird's Digestive System

Respiratory System Challenge

1. Why did God design a special respiratory system for birds? **God designed them with a special respiratory system that allows them to get a higher percentage of oxygen out of the air than most other animals. This is a very efficient system, which allows birds to fly for extended periods of time without tiring, and enables them to fly at high altitudes where there is less oxygen in the air.**

2. Do the lungs of birds expand and contract? **No**

What did we learn?

1. How does a bird "chew" its food without teeth? **God designed birds with an organ called a gizzard, which grinds the food up internally. In addition, some birds swallow small stones or pebbles that help to grind up the food as well.**

2. What purposes does the crop serve? **It holds the food so a bird can eat quickly. It then releases the food to be digested in a constant stream to provide a more constant source of energy.**

Taking it further

1. How is a bird's digestive system different from a human's digestive system? **A bird has a crop and a gizzard; humans do not. A bird does not have a large intestine. A bird's digestive system digests food much more quickly.**

2. How does a bird's digestive system help it to be a better flyer? **Because the food is digested quickly and efficiently, more energy is available for flying. Because the food is digested at a constant rate, a steady source of energy is provided for extended flying.**

God Designed the Bird's Digestive System Worksheet

See illustration in student manual.

11. Fish

What did we learn?

1. What makes fish different from other animals? **They live in the water, are cold-blooded, and have gills and scales**

2. How do fish breathe? **They get oxygen from the water using gills.**

3. Why do some sharks have to stay in motion? **They must have a constant flow of water over their gills to breathe, and the only way they can do that is to swim with their mouths open.**

4. What is the difference between warm-blooded and cold-blooded animals? **Warm-blooded animals regulate their body temperature—it stays the same regardless of the surrounding temperature. Cold-blooded animals cannot regulate their body temperature—it goes up and down with the surrounding temperature.**

Taking it further

1. Other than how they breathe, how are dolphins different from fish? **Dolphins are warm-blooded, give birth to live young, nurse their young, have hair, and do not have scales.**

2. How are dolphins like fish? **They live in the water, swim, have fins and a tail, and eat fish.**

Challenge: Designed for Speed

Placoid scales are flat tooth-shaped scales that are supported by tiny spines. They feel very rough if stroked from tail to head, but lay very flat when water moves from head to tail. These scales do not grow, so as a shark grows, more scales develop and fill in the gaps. Leptoid scales grow as the fish grows by adding material to the outer edge of each scale. They overlap each other from head to tail.

12. Fins & Other Fish Anatomy

What did we learn?

1. What is the purpose of a swim bladder? **It gives the fish buoyancy. When it fills with air, it makes the fish lighter than the water, allowing it to rise. When the air is released, the fish becomes heavier than the water and it sinks. This buoyancy keeps the fish floating without having to keep moving its fins.**

2. How did God design the fish to be such a good swimmer? **The shape of its body, its fins, and the mucus on its body all help it to be an efficient swimmer.**

Taking it further

1. How does mucus make a fish a more efficient swimmer? **Since mucus is slippery, it reduces friction so the fish does not have to work as hard to move through the water. To see how this works, put your hand under some running water and watch how the water flows. Then, rub a little cooking oil on your hand**

and repeat the test. The water flows more quickly over your oily hand because there is less friction.

2. How has man used the idea of a swim bladder in his inventions? **Submarines use air to help keep them afloat at the desired depth. Also, life rafts fill with air to help them float to the top of the water.**

3. What other function can fins have besides helping with swimming? **Fins can provide protection from predators. Fins can make it difficult for a predator to swallow a fish. In addition, some fins are shaped and colored to help provide camouflage.**

4. What similar design did God give to both fish and birds to help them get where they are going? **They both have rudder-like tails that help them steer, and bodies specially shaped for moving through their environments.**

Fish Fins Worksheet

See illustration in student manual.

13. Cartilaginous Fish

What did we learn?

1. How do cartilaginous fish differ from bony fish? **Their skeletons are made from cartilage instead of bone. Also, many of these fish do not have the typical torpedo-shaped body.**

2. Why is a lamprey called a parasite? **It does not eat prey. Instead, it attaches its mouth to a living animal, usually a fish, and sucks its blood for nutrients.**

3. Why can sharks and stingrays be dangerous to humans? **Sharks can attack with their sharp teeth and stingrays can sting with their tails.**

Taking it further

1. Why are shark babies born independent? **Like many other animals, sharks do not care for their young, so the babies must be able to care for themselves at birth. Many adult sharks will eat young sharks, so babies must avoid adults until they are large enough to defend themselves.**

2. What do you think is the shark's biggest natural enemy? **Other sharks**

Unit 3: Amphibians & Reptiles

14. Amphibians

What did we learn?

1. What are the characteristics that make amphibians unique? **They spend part of their lives in water breathing with gills, and part of their lives on land breathing with lungs. They are also cold-blooded, usually have smooth, moist skin, and lay eggs.**

2. How can you tell a frog from a toad? **In general, frogs have smooth, moist skin, while toads have dry, bumpy skin.**

3. How can you tell a salamander from a lizard? **Salamanders have smooth skin and lizards have dry scales on their skin. Also, salamanders go through a larval stage but lizards do not.**

Taking it further

1. What advantages do cold-blooded animals have over warm-blooded animals? **They don't have to eat as often and can usually survive a broader range of temperatures.**

2. What advantages do warm-blooded animals have over cold-blooded animals? **Cold-blooded animals' activities are more restricted by temperature extremes. A warm-blooded animal can still be quite active in very cold or very warm weather.**

3. Why are most people unfamiliar with caecilians? **Caecilians spend most of their time underground and live only in tropical rainforests, so most people never see them.**

15. Amphibian Metamorphosis

What did we learn?

1. Describe the stages an amphibian goes through in its lifecycle. **It begins as an egg, and then it hatches into a larva. In a frog, this is the tadpole stage. Then, it slowly changes into an adult. This is the metamorphosis stage in which lungs develop and gills disappear, and the creature changes its shape from a water dweller without legs to a land dweller with legs.**

2. What are gills? **They are special organs on the sides of water animals that extract oxygen from the water as water passes over or through them.**

3. What are lungs? **They are special organs that extract oxygen from the air as air passes through them.**

Taking it further

1. Does the amphibian life cycle represent molecules-to-man evolution? Why or why not? **NO! Evolution says that one kind of animal changes into another. A frog is still a frog even when it is a tadpole. A tadpole always changes into a frog. It does not grow up to be a bird or a mammal or even a salamander. It is always what God made it to be, even if its infant form is significantly different from its adult form.**

The World of Animals

Amphibian Lifecycle Worksheet

See illustrations in student manual.

16. Reptiles

What did we learn?

1. What makes reptiles different from amphibians? **Reptiles have scales and amphibians do not. Also, reptiles have lungs all their lives and do not go through metamorphosis.**
2. What are the four groups of reptiles? **Lizards, snakes, turtles, and crocodiles**

Taking it further

1. How do reptiles keep from overheating? **They stay in the shade or other cooler places during the hottest part of the day. Many sleep during the day and are only active at night.**
2. What would a reptile likely do if you dug it out of its winter hibernation spot? **It would appear dead. It would not move or eat. If you brought it inside and it warmed up, then it would seem to come alive, though it is actually alive even in its hibernating state.**

17. Snakes

What did we learn?

1. How are snakes different from other reptiles? **They have no legs.**
2. What are the three groups of snakes? **Constrictors, colubrids, and venomous snakes**
3. How is a snake's sense of smell different from that of most other animals? **It uses its tongue to collect scent particles, and then touches them to an organ called the Jacobson's organ inside its mouth.**
4. What is unique about how a snake eats? **It swallows its food whole and can eat something larger than its body diameter by unhooking its jaw and stretching its mouth very wide.**

Taking it further

1. How are small snakes different from worms? **Snakes have backbones, scales, and well-developed eyes. Worms do not have any of these. Also, snakes have much more complicated internal systems.**
2. If you see a snake in your yard, how do you know if it is dangerous? **You should learn to identify snakes using a guidebook or other resource. Unless you have your parent's permission, you should never approach a snake.**

18. Lizards

What did we learn?

1. List three ways a lizard might protect itself from a predator. **It could change its color, crawl into a crack in a rock and inflate its body, or break off its tail to escape.**
2. What do lizards eat? **Mostly insects; some eat plants and others eat dead animals.**

Taking it further

1. Horny lizards are short compared to many other lizards and are often called horny toads. What distinguishes a lizard from a toad? **Lizards are reptiles, toads are amphibians. Toads do not have scales but lizards do. Also, lizards do not have gills when they are young, nor do they experience metamorphosis, but toads do.**
2. Why might some people like having lizards around? **They eat insects and do not harm people.**
3. How does changing color protect a lizard? **It makes it hard for the predator to see it.**
4. What other reasons might cause a lizard to change colors? **To attract a mate or scare off competitors. Chameleons often use their coloring to communicate with other chameleons.**

19. Turtles & Crocodiles

What did we learn?

1. Where do turtles usually live? **In the water**
2. Where do tortoises usually live? **On land**
3. How does the mother crocodile carry her eggs to the water? **In her mouth**
4. Why can't you take a turtle out of its shell? **Its shell is part of its body.**
5. How do crocodiles stalk their prey? **They float in the water, wait for prey to approach, and then clamp their jaws around the prey and drag it under the water to drown it before eating it.**

Taking it further

1. Why might it be difficult to see a crocodile? **When it is floating in the water, it looks very much like a fallen**

log. This is how it tricks its prey into coming close enough to be eaten.

How Can You Tell Them Apart? Worksheet

Student drawings should be similar to pictures in student manual.

Unit 4: Arthropods

20. Invertebrates

What did we learn?

1. What are some differences between vertebrates and invertebrates? **The main difference is that vertebrates have backbones that protect their spinal cords. Invertebrates do not have backbones or spinal cords. Also, vertebrates have internal skeletons and invertebrates don't.**

2. What are the six categories of invertebrates? **Arthropods, mollusks, cnidarians, echinoderms, sponges, and worms**

Taking it Further

1. Why might we think that there are more vertebrates than invertebrates in the world? **We don't notice invertebrates as much as we do vertebrates. Invertebrates are usually small, a great many are microscopic, and so we just don't see them as often. Also, many invertebrates live in the water, so, again, we don't see them very often.**

21. Arthropods

What did we learn?

1. What do all arthropods have in common? **They are invertebrates [no backbone] with jointed feet, segmented bodies, and exoskeletons.**

2. What is the largest group of arthropods? **Insects, with over 1 million species**

Taking it Further

1. How are endoskeletons (internal) and exoskeletons (external) similar? **They both provide support and protection for the body. They help give the animal its form and shape.**

2. How are endoskeletons and exoskeletons different? **Endoskeletons are on the inside of the body and are usually made of bone or cartilage. Also, endoskeletons grow as the body grows. Exoskeletons are on the**

outside of the body and are made from chitin, a substance similar to starch. Exoskeletons do not grow with the animal and must be shed and replaced periodically as the body grows.

3. Why should you be cautious when hunting for arthropods? **Many arthropods are venomous, including some spiders, scorpions, and centipedes.**

Arthropod Pie Chart

1. _A_ Insects
2. _C_ Crustaceans
3. _B_ Arachnids
4. _E_ Centipedes
5. _D_ Millipedes

22. Insects

What did we learn?

1. What characteristics classify an animal as an insect? **Insects are invertebrates with jointed feet, 3 body parts [head, thorax, and abdomen], 6 legs, antennae, and usually have wings.**

2. How can insects be harmful to humans? **They can destroy crops and spread disease.**

3. How can insects be helpful? **Some insects eat other insects. For example, dragonflies eat mosquitoes, and ladybugs eat aphids. Many insects pollinate flowers. Other insects provide food for many other animals.**

Taking it Further

1. How might insects make noise? **Some insects make noise by flapping their wings. Others, like crickets, rub their legs together to make noises.**

23. Insect Metamorphosis

What did we learn?

1. What are the three stages of incomplete metamorphosis? **Egg, nymph, and adult**

2. What are the four stages of complete metamorphosis? **Egg, larva, chrysalis (or pupa), and adult**

Taking it Further

1. What must an adult insect look for when trying to find a place to lay her eggs? **The eggs must be laid on a plant that the larva can eat. A larva spends most**

of its time eating and cannot search for food, so food must be readily available.

Stages of Metamorphosis Worksheet

See student manual.

24. Arachnids

What did we learn?

1. How do arachnids differ from insects? **Arachnids have only two body parts (cephalothorax and abdomen), eight legs, no wings or antennae, and many spin webs.**

2. Why are ticks and mites called parasites? **They feed off of living hosts.**

Taking it further

1. Why don't spiders get caught in their own webs? **Only some of the web strands are sticky. The spider walks on the ones that are not sticky. Also, spiders secrete an oily substance that coats their feet and keeps them from sticking to their own webs.**

25. Crustaceans

What did we learn?

1. What do all crustaceans have in common? **They have jointed legs, exoskeletons, two body sections, two pairs of antennae, and two or more pairs of legs and gills.**

2. What are some ways that the crayfish is specially designed for its environment? **It has claws for defense and for eating. Its mouth is on the underside of its body, making it easier to eat food from the bottom of the river.**

Taking it further

1. Why might darting backward be a good defense for the crayfish? **It is unexpected and can confuse an enemy.**

2. At first glance, scorpions and crayfish (or crawdads) look a lot alike. How does a scorpion differ from a crayfish? **A scorpion lives on land and has eight legs, a stinger, and no antennae. Crayfish live in the water and have ten legs, antennae, and no stingers.**

3. How can something as large as a blue whale survive by eating only tiny crustaceans? **It eats lots and lots of them—up to 8,000 pounds (3,600 kg) per day!**

4. If you want to observe crustaceans, what equipment might you need? **Jar, net, microscope or magnifying glass, and a trap**

26. Myriapods

What did we learn?

1. How can you tell a centipede from a millipede? **Centipedes are usually smaller, flatter, and have longer antennae. Also, centipedes have 1 pair of legs per body segment while millipedes have 2 pairs per segment.**

2. What are the five groups of arthropods? **Insects, arachnids, crustaceans, centipedes, and millipedes**

3. What do all arthropods have in common? **They have jointed legs, exoskeletons, and 2 or more body segments.**

Taking it further

1. What are some common places you might find arthropods? **Nearly everywhere!**

2. Arthropods are supposed to live outside, but sometimes they get into our homes. What arthropods have you seen in your home? **Ants, flies, mosquitoes, spiders, etc.**

Unit 5: Other Invertebrates

27. Mollusks

What did we learn?

1. What are three groups of mollusks? **Bivalves, gastropods, and cephalopods**

2. What body structures do all mollusks have? **They all have soft bodies, a muscular foot, a hump for internal organs and a mantle that forms a shell in most species.**

3. How can you use a shell to help identify an animal? **The size, shape, and coloring of each shell are unique to its species. Some shells spiral counter-clockwise and others spiral clockwise. Some are two pieces and some are only one piece.**

Taking it further

1. How are pearls formed? **Any irritant that gets inside an oyster's shell is coated with a pearly substance over and over again. After a period of several years, it is large enough to be of value to people. To speed up this process, many oyster farmers now "seed" oysters by placing hard round objects that are nearly the size of a pearl inside oyster shells. After only a few months, these artificial pearls are ready for harvesting.**

28. Cnidarians

What did we learn?

1. What characteristics do all cnidarians share? **They have hollow bodies with stinging tentacles.**
2. What are the three most common cnidarians? **Jellyfish, corals, and sea anemones**

Taking it further

1. How do you think some creatures are able to live closely with jellyfish? **Accept reasonable answers. Some animals have a tough skin or exoskeleton that protects them from jellyfish stings. Others have a special coating on their skin that protects them.**
2. Why do you think an adult jellyfish is called a medusa? **The Medusa was a mythological creature with snakes for hair. A jellyfish, with all of its tentacles, resembles this creature.**
3. Jellyfish and coral sometimes have symbiotic relationships with other creatures. What other symbiotic relationships can you name? **Some birds eat insects off of cattle. This feeds the birds and helps the cattle stay healthy. Also, lichen, that green and yellow scaly-looking substance on rocks, is actually fungus and algae living in a symbiotic relationship. The algae have chlorophyll and produce the food, while the fungus provides water, nutrients, and protection. It is a beneficial relationship for both organisms.**

Challenge: Man O' War

Hydras can reproduce several different ways. They can reproduce by a process called budding where a new hydra forms from the side of the parent and then splits off. Hydras can also reproduce by sperm and eggs. Some species are able to produce both sperm and eggs from one animal. Other species have distinct male and female versions.

29. Echinoderms

What did we learn?

1. What are three common echinoderms? **Starfish (sea stars), sand dollars, and sea urchins**
2. What do echinoderms have in common? **They all have spiky skin and most have 5 body parts radiating from a central disk.**

Taking it further

1. Why would oyster and clam fishermen not want starfish in their oyster and clam beds? **Starfish can eat up to a dozen clams or oysters a day. This hurts the fishermen's business.**
2. What would happen if the fishermen caught and cut up the starfish and then threw them back? **The starfish would regenerate resulting in more starfish. This happened in one fishing village. The fishermen thought they were getting rid of the starfish by cutting them in half, but actually ended up making many more of them.**
3. What purpose might the spikes serve on echinoderms? **Most spikes are used for protection from predators.**

30. Sponges

Biomimetics Challenge

What three things can we learn from biomimetics? **First, we can see that even the "simplest" creatures are extremely complex. They have more to teach us than many people in the past have thought. Second, we see that God's design is better than man's design. Man can always learn from what God has created. Third, we see that God's creation declares His glory.**

What did we learn?

1. How does a sponge eat? **Nutrients are absorbed from the water as it passes through the body of the sponge.**
2. How does a sponge reproduce? **A sponge can reproduce by releasing eggs or a sponge can regenerate to form new sponges from pieces that are cut or broken off of the original sponge.**
3. Why is a sponge an animal and not a plant? **A sponge cannot produce its own food and it reproduces with eggs so it is an animal.**

Taking it further

1. Why can a sponge kill a coral colony? **It is immune to the poison darts of the coral.**
2. What uses are there for sponges? **They are sometimes used for cleaning, but mostly they are used for sponge painting and other artwork.**
3. Why are synthetic sponges more popular than real sponges? **They are much less expensive.**

31. Worms

What did we learn?

1. What kinds of worms are beneficial to man? **Segmented worms such as earthworms**
2. How are they beneficial? **They break down dead plant material, and can be used as fishing bait.**

3. What kinds of worms are harmful? **Most other kinds of worms are parasitic and thus are harmful to their hosts, whether they are human or animal hosts.**

Taking it further

1. How can you avoid parasitic worms? **Parasitic worms thrive in unsanitary conditions and are much more of a threat in undeveloped countries. Washing hands and raw vegetables, and cooking meat well will help you avoid most parasites.**

Unit 6: Simple Organisms

32. Kingdom Protista

What did we learn?

1. How are protists different from animals? **They consist of only one cell. Some contain chlorophyll.**
2. How are they the same? **Protists reproduce, eat, move, grow and need oxygen just like other animals. Also, protists have all the same cell parts as other animal cells.**

Taking it further

1. Why is a euglena a puzzle to scientists? **It has plant and animal characteristics.**
2. Why are single-celled creatures not as simple as you might expect? **Just because there is only one cell does not mean it is simple. Single-celled creatures perform very complex functions. Most protists are more complex than any cell in the human body because human cells are more specialized and protist cells must perform more functions. Even the smallest organism demonstrates God's marvelous powers of design, and refutes the idea that life evolved on its own.**

33. Kingdom Monera & Viruses

What did we learn?

1. How are bacteria similar to plants and animals? **They have cells, reproduce, and some can produce their own food.**
2. How are bacteria different from plant and animal cells? **They do not have a defined nucleus.**
3. How are viruses similar to plants and animals? **They have genetic information—DNA.**
4. How are viruses different? **They do not reproduce on their own. They do not eat or grow in a normal sort of way.**

Taking it further

1. Answer the following questions to test if a virus is alive.

   Does it have cells? **No.**

   Can it reproduce? **Only with the help of a host cell.**

   Is it growing? **Not in the ordinary sense of the word.**

   Does it move or respond to its environment? **Yes.**

   Does it need food and water? **It needs host cells that use food and water. It is unclear if the viruses use these things directly.**

   Does it have respiration? **No.**

   Is it alive? **No, it does not have all of the requirements for biological life.**

2. How can use of antibiotics be bad? **Antibiotics kill bacteria, but they cannot distinguish between good and bad bacteria. Overuse of antibiotics can kill too many of the good bacteria in your intestines and cause problems. Also, antibiotics kill most of the bad bacteria but some are resistant and do not die. These bacteria are the ones that survive and reproduce. The next generation of bacteria is not as easily killed by the antibiotics. Doctors are beginning to see diseases that used to respond to certain antibiotics no longer respond and must now be treated with stronger medicines. So we need to carefully use antibiotics when necessary, but not overuse them or use them incorrectly.**

34. Animal Notebook

What did we learn?

1. What do all animals have in common? **They are alive, they reproduce, they do not make their own food, they can move about during at least part of their life.**
2. What is the difference between vertebrates and invertebrates? **Vertebrates have a backbone and invertebrates do not.**
3. What sets protists apart from all the other animals? **They are single-celled creatures. Some, like the euglena, can make their own food.**

Taking it further

1. What are some of the greatest or most interesting things you learned from your study of the world of animals? **Answers will vary.**
2. What would you like to learn more about? **Check out books from the library to learn more.**

The World of Plants — Quiz Answer Keys

Quiz 1. Introduction to Life Science
Lessons 1–4

Mark each statement as either True or False.
1. _T_ All living creatures have cells.
2. _F_ Plants do not need oxygen.
3. _T_ Growth and change can be signs of life.
4. _F_ Nonliving things absorb nutrients.
5. _F_ Plants cannot move, so they are not alive.
6. _T_ A kingdom is a way to group things together by similar characteristics.
7. _F_ Plants and protists are the two main kingdoms of living things.
8. _F_ Plants and animals both have chlorophyll.
9. _T_ Vacuoles store food inside of cells.
10. _T_ The nucleus is the control center of a cell.

Short answer:
11. Name three differences between plant cells and animal cells. **Shape of the cells—plant cells are usually rectangular, animal cells are usually round; plant cells have chlorophyll, animal cells do not; plant cells have a cell wall, animal cells do not; only plant cells perform photosynthesis.**
12. Describe how to tell if something is alive. **It eats, "breathes," grows, reproduces, moves/responds to its environment, and has cells.**

Challenge questions
Fill in the blanks using the terms below. Not all words are used.
1. The law of _biogenesis_ states that life always comes from life.
2. _Chemical evolution_ says that life originally came from nonliving chemicals.
3. During _mitosis_ a cell divides into two identical cells.
4. A _conifer or gymnosperm_ is a type of plant that produces seeds in cones.
5. A _Ginkgo_ tree is sometimes called a living fossil.
6. _Angiosperms_ produce seeds that are enclosed in fruit.
7. The belief that life springs up from its environment is called _spontaneous generation_.
8. During _metaphase_ the chromosomes in a cell line up in the middle.
9. During _anaphase_ the duplicate chromosomes are pulled apart.
10. _Cytokinesis_ occurs when the cytoplasm in a cell is divided.

Quiz 2. Flowering Plants & Seeds
Lessons 5–10

Match the term with its definition.
1. _A_ Monocot
2. _D_ Dicot
3. _B_ Cotyledon
4. _F_ Hilum
5. _I_ Plumule
6. _G_ Radicle
7. _C_ Seed Coat
8. _E_ Deciduous
9. _H_ Evergreen
10. _J_ Angiosperms

Answer Yes or No. Would a seed germinate if placed in each of the following conditions? If no, explain what is missing.
11. A seed planted in a garden in the spring time and watered every day. **Yes—All necessary conditions are present.**
12. A seed planted in a desert. **Probably not—There is probably not enough water.**
13. A seed planted in the dust on the moon. **No—There is not enough oxygen, moisture, or heat on the moon.**
14. A seed in an envelope at the store. **No—There is not enough moisture in the envelope.**

15. A seed in a moist paper towel in a sunny window. **Yes—All necessary conditions are present.**

Short answer:

16. Name the four organs of a plant. **Roots, Stem, Leaves, Flowers**

Challenge questions

Mark each statement as either True or False.

1. _T_ Many plants can be used to make medicines.
2. _F_ All grass is alike.
3. _T_ Rye, wheat, and oats are all grasses.
4. _T_ Many trees have distinctive crowns.
5. _T_ A pine tree has a triangular growth habit.
6. _F_ Pruning will not affect a tree's growth habit.
7. _F_ External seed dormancy depends on temperature.
8. _T_ Some commercial growers use sulfuric acid to scarify seeds.
9. _T_ Stratification of seeds can occur in a refrigerator.
10. _F_ Seeds with double dormancy can experience scarification and stratification in either order and still germinate.

Quiz 3. Roots & Stems

Lessons 11–15

Choose the best answer for each statement or question.

1. _B_ Which is not a function of the roots of a plant?
2. _A_ Which is not considered a plant organ?
3. _C_ A plant with this type of roots is most likely to live where it is dry.
4. _D_ A plant with this type of roots is most likely to live in a tree.
5. _A_ You would be most successful planting plants with these roots on a steep hill.
6. _B_ A plant with this kind of roots will be more successful in a very wet area.
7. _A_ Which is not a function of the stem?

8. _C_ What causes water to move upward in a plant?
9. _D_ What is a new stem called?
10. _C_ Where do leaves connect to the stem?
11. _B_ Which kind of cells are not found inside a stem?
12. _A_ Which cells carry water up a stem?
13. _C_ Which cells carry food down a stem?
14. _B_ Which cells protect a young stem?
15. _D_ Which cells protect mature stems?

Short answer:

16. How do parasitic roots "steal" nutrients from nearby plants? **By sending out special shoots called haustoria which tap into the adjacent plant's roots to "steal" nutrients from the other plant.**

Challenge questions

Choose the best answer for each question.

1. _B_ How does primary growth change the root?
2. _A_ Where does primary growth occur in a root?
3. _B_ How is osmosis different from diffusion?
4. _D_ What role does capillarity play in plants?
5. _C_ Which type of branching results in a wide, low tree or shrub?

Quiz 4. Leaves

Lessons 16–20

1. What is the purpose of the stomata in leaves? **To allow carbon dioxide in and oxygen out**
2. For each leaf below, describe its vein arrangement (palmate or pinnate) and identify it if you can.
   **A. Palmate—Maple
   B. Pinnate—Oak
   C. Pinnate—Ash
   D. Pinnate—Holly**

Short answer:

3. Which leaf above appears to be a compound leaf? **C**
4. What kind of leaf arrangement does plant D have? **Alternate**
5. How do leaves follow the sun? **The tips of the leaves detect the light and send out a chemical which makes**

the cells on the shady side get longer, thus turning the leaf toward the sun.

6. What makes leaves green? **The chlorophyll in the chloroplasts**

7. Identify the "ingredients" (beginning materials) and the "products" (ending materials) of photosynthesis. Ingredients: **Light, water, carbon dioxide, and chlorophyll;** Products: **Glucose/sugar and oxygen**

Challenge questions

Short answer:

1. For each leaf above describe the leaf margin. **A. Lobed   B. Lobed   C. Toothed   D. Toothed**

2. Name two pigments that could be found in leaves. **Chlorophyll, carotene, xanthophyll, anthocyanin, etc.**

3. What is the chemical formula for photosynthesis? $6\,CO_2 + 6\,H_2O + light = C_6H_{12}O_6 + 6\,O_2$

4. What is the purpose of a bract? **To attract pollinators to the flowers**

5. What is one purpose of a succulent leaf? **To store water, to perform photosynthesis**

Quiz 5. Flowers & Fruits

Lessons 21–25

1. Identify the parts of the flower below:
   a. **Pollen**, b. **Ovary**, c. **Ovule**, d. **Pistil**, e. **Petals**, f. **Stamen**, g. **Sepal**

Short answer:

2. Which part of the flower is considered the male part? **Stamen**

3. Which part of the flower is considered the female part? **Pistil**

4. How is a simple fruit different from an aggregate fruit? **Simple fruit forms one fruit from one flower with one pistil. Aggregate fruit forms one fruit from one flower with multiple pistils.**

5. Describe the process of pollination. **Pollen is removed from a stamen, usually by a pollinator like a bee or other insect. It is then deposited on the pistil of another flower. A pollen tube grows down into the ovary until it reaches the ovule. Fertilization takes place and the ovule becomes a seed.**

6. List two ways that pollen can be transferred from one flower to another. **Possible answers include: by an animal or other pollinator, by wind, rain, or by a person.**

7. How long does it take for a biennial to complete its lifecycle? **2 years or 2 growing seasons**

Challenge questions

Mark each statement as either True or False.

1. _F_ Composite flowers have only one flower per stalk.
2. _T_ Ray flowers often look like petals.
3. _T_ Disk flowers produce hundreds of seeds.
4. _F_ Flowers do not need to attract pollinators.
5. _T_ Some nectar guides can normally only be seen by insects.
6. _T_ Some flowers smell bad to attract flies as pollinators.
7. _T_ Succulent fruits are simple fruits.
8. _T_ An olive is considered a fruit.
9. _F_ Peanuts are nuts.
10. _F_ Apples are berries from a biological definition.

Quiz 6. Unusual Plants

Lessons 26–34

Match the term with its definition.

1. _C_ Plants that eat insects
2. _B_ Plants that "steal" nutrients from other plants
3. _D_ Plants that grow on other plants without harming them
4. _A_ Response of plants to gravity (roots go down, stems go up)
5. _E_ Tendency for roots to grow toward water
6. _G_ Ability of leaves to turn toward sunlight
7. _F_ Plant designed to store available water in dry conditions
8. _J_ Plant reproduction using a part of the plant (not seeds)
9. _H_ Runners from a strawberry plant

10. _I_ Special stems that grow underground for reproduction

Short answer:

11. What plant organ is missing in ferns? **Flowers**

12. How do both mosses and ferns reproduce? **Spores**

13. What two plant organs are missing in mosses? **Flowers and true roots**

14. What substance do algae have in common with plants? **Chlorophyll**

15. What is the name of the group that contains yeast and mushrooms? **Fungi**

Challenge questions

Short answer: (Accept reasonable answers for all challenge questions.)

1. Give an example of positive tropism. **Hydrotropism, phototropism, and geotropism for roots**

2. Give an example of negative tropism. **Geotropism for stems, thermotropism for curling leaves, and thigmotropism for roots**

3. Explain how a cobra lily traps insects. **It attracts insects to its pitcher with nectar, then as the insect tries to find the exit it hits the top of the pitcher and falls inside.**

4. Explain how succulents are designed to survive dry periods. **They can store large amounts of water in their stems and/or leaves.**

5. Why is grafting a form of cloning? **The resulting plant has identical DNA to the parent plant.**

The Human Body — Quiz Answer Keys

Quiz 1. Body Overview
Lessons 1–3

Fill in the blanks with the correct term.
1. Many cells working together are called a _tissue_.
2. _Red blood_ cells carry oxygen to the body.
3. _Muscle_ cells stretch and contract to allow for movement.
4. The _cell membrane_ acts like the skin of a cell.
5. The _nucleus_ is the brain or control center of a cell.
6. Vacuoles are where cells store _food (or other nutrients, or waste)_.
7. Mitochondria break down food to provide _energy_ for the cell.
8. Humans were created in _God's_ image.
9. The _skeletal_ system provides strength for the body.
10. _Nerve_ cells can be over a yard long.
11. Nutrients are provided to the body through the _circulatory_ system.
12. _Skin_ protects the body from harmful substances outside the body. **(also accept white blood cells)**
13. _Bone_ cells have a criss-cross shape.
14. Several tissues working together are called an _organ_.
15. A system is made up of _cells, tissues,_ and _organs_ all working together to perform a specific function.
16. Explain the function of red blood cells and white blood cells. **Red blood cells carry oxygen to and carbon dioxide away from the cells of the body. White blood cells eliminate invading germs.**

Challenge questions
Fill in the blanks with the correct term.
1. The _endocrine_ system produces chemical messengers to control body functions.
2. Waste products are removed from the body by the _kidneys or excretory system_.
3. A mother carries the unborn baby in her _womb or uterus_.
4. Bones are _connective_ tissue.
5. _Epithelial_ tissue covers the inside of your stomach.

Quiz 2. Bones & Muscles
Lessons 4–10

Place the letter for the correct bone type next to each description below.
1. _B_ Makes new blood cells
2. _A_ Found in hands and feet
3. _B_ Supports most of your weight
4. _B_ Found in arms and legs
5. _D_ Found in face
6. _C_ Gives protection to organs
7. _D_ Vertebrae
8. _A_ Gives flexibility in hands
9. _C_ Ribs and skull
10. _C_ Shoulder blades

Mark each statement as either True or False.
11. _F_ Muscles stretch and contract individually.
12. _T_ Muscles can be damaged by tearing.
13. _T_ Using muscles makes them stronger.
14. _T_ Approximately 40% of your body weight is from muscles.
15. _F_ What you eat does not affect your bones and muscles.

Challenge questions
Mark each statement as either True or False.
1. _T_ The arm and leg bones are part of the appendicular skeleton.
2. _F_ The axial skeleton primarily provides form and strength.
3. _T_ Blood-clotting cells are some of the first cells at the site of a fracture.
4. _F_ Broken bones are usually weaker after they heal than before the break.

5. _T_ It can take weeks for a bone to fully heal.
6. _T_ Joints are designed to keeps bones in place and to move freely.
7. _T_ Individual muscles cells each contract to make a muscle contract.
8. _T_ Muscles help blood move through the body.
9. _F_ The heart is made up of skeletal muscle tissue.
10. _T_ Only hands and feet have friction skin.

Quiz 3. Nerves & Senses

Lessons 11–18

1. Name the five senses your brain uses to collect information about the outside world.

   Sight Taste Touch
   Hearing Smell

Match each part of the brain with its function.

2. _D_ Cerebellum
3. _F_ Brain stem
4. _E_ Spinal cord
5. _C_ Cerebrum
6. _B_ Hippocampus
7. _A_ Pituitary gland

Mark each statement as either True or False.

8. _T_ You can improve your intelligence by exercising your brain.
9. _T_ Smells can bring back memories.
10. _F_ You don't need to wear a bike helmet when you ride your bike.
11. _F_ The left side of the brain controls the left side of the body.
12. _T_ What you eat affects your brain.
13. _F_ Reflexes are slower than normal signals to the brain.
14. _T_ Your brain fills in for the blind spot in your eye.
15. _T_ The louder the sound is the higher its amplitude.

Challenge questions

Choose the best answer for each question below.

1. _B_ Which nervous system is the only one capable of higher level complex thought?
2. _C_ Which neurons process and generate signals?
3. _A_ What is the function of myelin?
4. _D_ What liquid is found in the middle of the eye?
5. _C_ What part of the ear controls balance?

Quiz 4. Digestive System

Lessons 19–23

1. Name the six major parts of the digestive system.
   **Teeth Tongue Esophagus Stomach
   Small intestine Large intestine**

Match the type of tooth with its function.

2. _B_ Incisors
3. _A_ Canines
4. _C_ Bicuspids
5. _C_ Molars

Short answer:

6. Why is it important to take good care of your teeth? **So you don't get cavities or gum disease. To keep your teeth healthy.**
7. How do you take good care of your teeth? **Brush and floss regularly and visit your dentist regularly.**
8. Why is important to eat healthy foods? **To have energy to do the things you want to do and to keep your body healthy.**
9. Name the five different food groups you learned about. **Bread/grains, Fruits, Vegetables, Dairy, Meat/beans**
10. Name the three forms of energy in food. **Carbohydrates, proteins, fat**
11. What two other important types of compounds do we get from our food? **Vitamins and minerals**

Challenge questions

Mark each statement as either True or False.

1. _T_ Enzymes play a crucial role in digestion.
2. _F_ The gall bladder stores gastric juice.
3. _T_ Pancreatic juice helps break down fats.
4. _T_ Dentin gives the tooth its general size and shape.
5. _F_ Enamel is one of the softest substances in the body.
6. _T_ Orthodontics is the area of dentistry that corrects the alignment of teeth.
7. _T_ A banana has fewer calories than a cup of french fries.
8. _T_ Many diseases can be prevented by eating the right foods.
9. _F_ Rickets can cause bleeding of the gums.
10. _T_ Eating green leafy vegetables can help prevent anemia.

Challenge questions

Short answer:

1. What is the difference between systolic and diastolic blood pressure? **Systolic blood pressure is the pressure in the blood vessels when the heart is contracting, diastolic is the pressure when the heart is resting.**
2. List two ways that a person might lower his/her blood pressure. **Change in diet, exercise, and medication**
3. Describe how blood moves through the heart. **Blood enters the right atrium then fills the right ventricle. It is then pumped out to the lungs. Blood returning from the lungs enters the left atrium and then fills the left ventricle. It is then pumped out of the heart to the rest of the body.**
4. Which blood type or types can donate to someone with A positive blood? **O positive, A positive, O negative, A negative.**
5. Explain the difference between external and internal respiration. **External respiration takes place in the lungs, internal respiration takes place between the blood cells and the tissue cells.**

Quiz 5. Heart & Lungs

Lessons 24–28

Fill in the blank with the part of blood that does each job below.

1. **Red blood cells** transport oxygen and carbon dioxide.
2. **White blood cells** surround and eliminate germs.
3. **Plasma** carries the cells around the body.
4. **Platelets** repair breaks in the blood vessels.

Match the term with its definition.

5. _C_ Diaphragm
6. _A_ Carbon dioxide
7. _D_ Oxygen
8. _B_ Bronchi
9. _E_ Trachea
10. List two ways to keep your lungs healthy: **Exercise, don't smoke, and avoid exposure to people who are ill.**

Quiz 6. Skin & Immunity

Lessons 29–33

Choose the best answer for each question.

1. _B_ Which of the following is not a part of your skin?
2. _D_ Which cells are found on the outermost part of the epidermis?
3. _C_ Where is friction skin found on your body?
4. _A_ Which of the following is not a function of the skin?
5. _D_ Which people have the same fingerprints?
6. _B_ Which organ provides a barrier against disease?
7. _A_ Which of the following are produced by white blood cells?
8. _C_ Which of the following does not help filter germs from your body?
9. _A_ How are physical traits passed on from parent to child?
10. _C_ Which of the following grows from a follicle?

Challenge questions

Match the term with its definition.

1. _B_ Arrector pili muscles
2. _A_ Albinism
3. _C_ Double helix
4. _E_ Base pair
5. _G_ Thymine
6. _J_ Cytosine
7. _H_ Chromosome
8. _D_ Deoxyribose
9. _F_ Adenine
10. _I_ Mutation

The World of Animals — Quiz Answer Keys

Quiz 1. Mammals

Lessons 1–7

Short answer:

1. What are the two main groups of animals? **Vertebrates and invertebrates**

2. What are the five major groups of vertebrates? **Mammals, birds, amphibians, reptiles, fish**

3. What are five common characteristics of mammals? **Warm-blooded, breathe air with lungs, fur/hair, live birth, nurse young**

4. What makes a marsupial different from other mammals? **Has a pouch**

5. What makes a vertebrate unique? **Has a spinal cord ending in a brain, backbone**

Mark each statement as either True or False.

6. _F_ Animals can produce their own food.
7. _F_ Dolphins are large fish.
8. _T_ Marsupials give birth to tiny live babies.
9. _F_ Baleen whales have large teeth.
10. _F_ The elephant is the largest animal in the world. **Note: Elephants are the largest land animal, but blue whales are the largest animal on earth.**
11. _T_ Monkeys have tails, but apes do not.
12. _T_ Marsupials live primarily in Australia and Tasmania.
13. _T_ Some marsupials are meat-eating animals.
14. _T_ A lemur is a primate.
15. _F_ Primates have eyes on the sides of their heads.

Challenge questions

1. Draw a foot of a mammal to represent each of the following stances:

    Unguligrade: **Any animals with a hoof—horse, deer, or elephant.**

    Digitigrade: **Any animal that walks on its toes—dog, cat, wolf, or fox.**

    Plantigrade: **Any animal that walks on flat feet—rabbit, skunk, opossum, monkey, or bear.**

2. Match the parts of a ruminant's digestive system with its definition.

    _A_ Rumen _E_ Abomasum

    _C_ Cud _B_ Reticulum _D_ Omasum

3. List three special design features that God gave whales. **Blowhole separate from mouth, ability to regulate temperature with fins, ability to regulate blood pressure/body pressure when diving, ability to shut down unnecessary functions while feeding, echolocation, ability to expand throat when eating**

4. List three special design features that God gave koalas. **Ability to eat and digest leaves, special teeth, bacteria in stomach, friction skin on feet for climbing, thumbs for grasping, pouch for babies**

5. Explain why an ape doing sign language does not necessarily support the evolution of man from apes. **Many animals have some level of intelligence, but none are close to humans. More importantly, humans have a spirit that relates to God. Sign language does not necessarily demonstrate humanity.**

Quiz 2. Birds & Fish

Lessons 8–13

Short answer:

1. Look at each picture of birds' feet below. Below each picture write the term you think is most appropriate: **A. Water B. Bird of prey C. Perching D. Runner**

2. Look at each picture of birds' beaks below. Next to each picture write what you think that bird is likely to eat: **A. Seeds B. Water plants C. Other animals D. Nectar**

3. List three ways that birds were specially designed for flight. **Strong breast muscles, stiff backbone, hollow bones, efficient digestive and respiratory systems, and feathers**

4. List two special design features of a bird's digestive system. **It is very fast and efficient, crop releases food at a constant rate, and gizzard grinds food**

5. Name three kinds of fins found on most fish. **Pectoral, pelvic, dorsal, anal, caudal**

6. Name two kinds of cartilaginous fish. **Sharks, rays, hagfish, and lampreys**

Challenge questions

Mark each statement as either True or False.

1. _T_ Animals can adapt to changes in their surroundings.
2. _F_ Birds evolved from reptiles.
3. _F_ Birds and reptiles are both cold-blooded animals.
4. _T_ Scales are very different from feathers.
5. _T_ There can be great variety among species.
6. _T_ Penguins only live in the southern hemisphere.
7. _T_ Birds have a very efficient respiratory system.
8. _F_ Birds have a bellows type of respiratory system.
9. _T_ Fish scales fit together smoothly from head to tail.
10. _F_ Fish have very small olfactory lobes compared to their brain size.

Quiz 3. Amphibians & Reptiles

Lessons 14–19

1. What defines an animal as a vertebrate? **Vertebrates have backbones with a spine along the back ending in a brain.**

Place the letters of the characteristics that apply next to each animal group.

2. Mammals: **A, C, G, J, L (H for some)**
3. Birds: **A, E, H, J, M**
4. Fish: **B, D, H, K, N (G for a few)**
5. Reptiles: **B, D, H, J**
6. Amphibians: **B, F, H, I, J, K**

Challenge questions

Fill in the blank with the correct term from below.

1. Amphibians communicate primarily by _**sound**_.
2. Male frogs have inflatable _**air sacs**_ for communication.
3. Each frog species communicates on a different _**frequency**_.
4. The Surinam toad presses eggs into the mother's _**back**_.
5. The midwife toad carries eggs strapped to its _**legs**_.
6. The mouth brooding frog carries its tadpoles in its _**mouth/air sacs**_.
7. A _**triceratops**_ is a ceratopsian dinosaur.
8. An _**allosaurus**_ is a theropod dinosaur.
9. An _**apatosaurus**_ is a sauropod dinosaur.
10. Marine iguanas are well adapted to life in and near the _**water**_.

Quiz 4. Arthropods

Lessons 20–26

Write Yes if the creature below is an arthropod, write No if it is not.

1. _**Yes**_ ant
2. _**Yes**_ tick
3. _**No**_ trout
4. _**Yes**_ spider
5. _**Yes**_ scorpion
6. _**Yes**_ crab
7. _**Yes**_ cricket
8. _**Yes**_ butterfly
9. _**No**_ clam
10. _**No**_ mouse
11. _**Yes**_ centipede
12. _**No**_ snail
13. _**Yes**_ roly-poly
14. _**Yes**_ crawdad
15. _**No**_ starfish
16. _**No**_ lizard
17. What are the four stages of complete metamorphosis for an insect? **egg, larva, chrysalis (or pupa), adult**

Fill in the blanks with the appropriate numbers.

18. An insect has _**3**_ body parts, _**6**_ legs, _**2**_ antennae and _**2 or 4**_ wings.

19. A spider has _2_ body parts, _8_ legs, _0_ antennae and _0_ wings.
20. A centipede has _1_ pair(s) of legs per body segment and a millipede has _2_ pair(s) of legs per body segment.

Challenge questions

Short answer:

1. Explain the purpose of an arthropod's exoskeleton. **The exoskeleton provides protection, keeps the animal from drying out, gives it form.**
2. What is the main ingredient in an exoskeleton? **Starch or chitin**
3. Identify which segment (head, thorax, abdomen) is primarily responsible for each of the following functions in insects.

   Locomotion: **Thorax**

   Internal functions: **Abdomen**

   Sensory input: **Head**
4. What are two purposes of bioluminescence in fireflies? **Protection from enemies, finding a mate**
5. Do male or female tarantulas live longer? **Female**
6. What do tarantulas usually eat? **Grasshoppers, crickets, and other small animals**
7. List three types of symmetry commonly found among animals. **Bilateral, radial, spherical**
8. What does it mean if an animal is asymmetrical? **It has no symmetry; it cannot be divided into similar halves**
9. How does a millipede protect itself from predators? **It coils up, produces smelly or caustic liquid**
10. Which is more dangerous, a centipede or a millipede? **A centipede is venomous.**

Quiz 5. Other Invertebrates
Lessons 27–31

Mark each statement as either True or False.

1. _F_ All mollusks have visible shells.
2. _T_ A bivalve has two parts to its shell.
3. _T_ You can identify a mollusk by the shape of its shell.
4. _F_ The octopus is considered one of the least intelligent invertebrates.
5. _T_ Cnidarians usually experience a polyp stage sometime in their lifecycle.
6. _T_ Coral and algae have a symbiotic relationship.
7. _F_ Echinoderms are usually very dark colors.
8. _T_ Several invertebrates have the ability to regenerate.
9. _F_ Echinoderms have smooth skin.
10. _T_ A sponge is one of the simplest invertebrates.
11. _T_ Sponges can reproduce by eggs.
12. _F_ All worms are harmful to humans.

Short answer:

1. What do jellyfish, coral and sea anemones have in common? **They all have hollow bodies and stinging tentacles at least during part of their lifecycles.**
2. Name three groups of worms. **Segmented, flat, round.**
3. What part of the mollusk secretes its shell? **Mantle**
4. Which kind of mollusk has only one part to its shell? **Gastropod**
5. What is an adult jellyfish called? **Medusa**

Challenge questions

Short answer:

1. Explain how cephalopods move. **Jet propulsion— suck in water then force it out the back**
2. How can a nautilus remain buoyant as its shell gets bigger and heavier? **It fills inner chambers with gas/air.**
3. What is a siphonophore? **A collection of cnidarians living together in a symbiotic relationship.**
4. Name a common siphonophore. **Portuguese Man-of-War, by-the-wind sailor**
5. What is a madreporite in a starfish? **The openings that allow water into the starfish's water vascular system.**
6. What technology is being improved by the study of the Venus Flower Basket sponge? **Fiber optics**

7. What is the name of the process that provides food for tubeworms? **Chemosynthesis**

8. What is a symbiotic relationship? Describe two examples of cnidarians that have symbiotic relationships. **A symbiotic relationship is one where both benefit in some way. Accept reasonable answers noting the benefits to both. For example: corals with algae living inside them – algae produces food for the coral, while coral protects the algae; some small fish protected from larger fish by nearby jellyfish – jellyfish stings the larger fish and both it and the small fish eat the larger one.**

Short Answer

9. Many protists are parasitic and live in water. Name two of the serious diseases they can cause. **Answers should include two of the following: Malaria, African sleeping sickness, and amoebic dysentery.**

10. How do protists generally reproduce? **Protists generally reproduce by some sort of division where one cell divides to form two new cells.**

Quiz 6. Simple Organisms

Lessons 32–34

Match the parts of a cell to its function.

1. _E_ Nucleus
2. _C_ Cell membrane
3. _B_ Cytoplasm
4. _A_ Mitochondria
5. _D_ Vacuole

Match the single-celled organism with its description.

6. _C_ Flagellate
7. _A_ Sarcodine
8. _B_ Ciliate
9. _E_ Bacteria
10. _D_ Virus

Challenge questions

Mark each statement as either True or False.

1. _F_ Sporozoans have a very simple lifecycle.
2. _T_ Sporozoans reproduce asexually and sexually.
3. _T_ Sporozoans are parasites.
4. _T_ Plasmodium is a dangerous protist.
5. _F_ Antibiotic-resistant bacteria prove evolution.
6. _F_ Survival of the fittest is the same as evolution.
7. _T_ Fossilized bacteria are very similar to modern bacteria.
8. _T_ Bacteria support biblical creation.

The World of Plants — Final Exam Answer Key
Lessons 1–34

Define each of the following terms.

1. Geotropism: **Response of plant to gravity; causes roots to grow down and stems to grow up**
2. Angiosperm: **Plant that reproduces with flowers**
3. Phototropism: **Response of plants to light; leaves turn toward the sun**
4. Photosynthesis: **Process by which sunlight, chlorophyll, carbon dioxide, and water are turned into sugar and oxygen**
5. Pollination: **Process by which pollen is transferred from one flower to another to cause the fertilization of the ovule, thereby creating seeds**
6. Chlorophyll: **The green substance in plant cells that is used to perform photosynthesis**
7. Ovule: **The egg or unfertilized seed found in the ovary of the flower.**
8. Pistil: **The female part of the flower that produces ovules**
9. Stamen: **The male part of the flower that produces pollen**
10. Xylem and phloem: **The tubes that carry nutrients, food, and water throughout the plant.**

Choose the best answer for each question.

11. _C_ Which of the following is not an organ of flowering plants?
12. _D_ Which kind or kinds of creatures get nourishment from grasses? (People and birds eat the seeds.)
13. _A_ Which is not a common use for the wood of a tree?
14. _B_ Which tree is a deciduous tree?
15. _C_ Which organ is primarily used to absorb minerals from the ground?

Fill in the blank with the correct term.

16. Root growth primarily occurs at the _**root tip or root cap**_.
17. The two types of root systems are _**fibrous**_ and _**taproot**_.
18. The shape of most monocot plants' leaves is _**long and thin like grass**_.
19. Ferns reproduce by _**spores**_ on their fronds.
20. Algae are similar to plants because they contain _**chlorophyll**_.

Mark each statement as either True or False.

21. _F_ Plants with red leaves have no chlorophyll.
22. _T_ Trees can be identified by their leaves.
23. _F_ Coniferous trees do not have leaves.
24. _F_ The scent of a flower has no purpose.
25. _T_ Ferns are not flowering plants.
26. _T_ Algae is an important organism.
27. _T_ Sepals might be confused with leaves.
28. _T_ Pollination must take place for seeds to form.
29. _T_ Mosses reproduce by spores and not seeds.
30. _T_ Photosynthesis cannot take place without chlorophyll.

Challenge questions

Match the term with its definition.

1. _E_ Law of biogenesis
2. _C_ Meiosis
3. _A_ Spontaneous generation
4. _D_ Scarification
5. _F_ Stratification
6. _B_ Seed dormancy
7. _J_ Primary growth
8. _I_ Secondary growth
9. _G_ Osmosis
10. _H_ Toothed leaf margin
11. _K_ Lobed leaf margin

Mark each statement as either True or False.

12. _F_ Ephemeral plants grow slowly.

13. _T_ Composite flowers are really hundreds of flowers grouped together.
14. _F_ Legumes have hard outer shells.
15. _T_ Pomes have papery inner cores.
16. _T_ Chemotropism aids in pollination.
17. _T_ Filament algae is very common.
18. _F_ Most commercial fruit trees are grown from seeds.
19. _F_ Tendrils have negative tropism.
20. _T_ Rootstock is important for grafting.

The Human Body — Final Exam Answer Key

Lessons 1–34

Choose the best answer for each question.

1. _A_ Which of the following are the three main types of fingerprints?
2. _C_ What kind of skin is on the palms of your hands and bottoms of your feet?
3. _B_ Which are not found in skin?
4. _B_ Which is determined by genetics?
5. _D_ Your hip is which type of joint?
6. _A_ What causes a muscle to contract?
7. _C_ Which is not a part of the nervous system?
8. _B_ What is the main function of the hippocampus in the brain?
9. _C_ Which should you eat sparingly for good health?

Match the name of the system to its function.

10. _D_ Circulatory system
11. _A_ Digestive system
12. _E_ Skin
13. _F_ Nervous system
14. _B_ Muscular system
15. _G_ Respiratory system
16. _C_ Skeletal system

Mark each statement as either True or False.

17. _T_ Long bones produce red blood cells.
18. _F_ Your funny bone is a bone in your arm.
19. _T_ It is important to drink plenty of water every day.
20. _T_ You will get most of the vitamins you need if you eat a variety of foods.
21. _F_ White blood cells stop a cut from bleeding too much.
22. _T_ A human heart has four chambers.
23. _T_ Your lungs are filled with branching tubes that end in alveoli.
24. _F_ Your brain is the largest organ in your body.
25. _T_ Skin contains pain and temperature receptors.

Challenge questions

Mark each statement as either True or False.

1. _T_ The endocrine system produces hormones.
2. _F_ The excretory system removes nutrients.
3. _T_ The reproduction system was designed to create new life.
4. _F_ Epithelial tissue moves muscles.
5. _T_ Connective tissue holds the body together.
6. _T_ The skeletal system is divided into axial and appendicular parts.
7. _T_ Only humans have the ability to perform complex reasoning.
8. _F_ Cartilage holds bones in place.
9. _F_ Muscles only move bones.
10. _T_ Reflexes are faster than other nervous system responses.
11. _T_ Dendrites receive input.
12. _F_ The Braille alphabet was developed to allow deaf people to communicate.
13. _F_ Rods and cones allow you to eat ice cream.
14. _F_ The cochlea is part of the middle ear.
15. _T_ Enzymes speed up digestion.

Short answer:

16. Name five bones in your body. **Accept reasonable answers.**
17. Name five muscles in your body. **Accept reasonable answers.**
18. Name four functions of the brain. **Controlling the body, moving the body, thinking, remembering, controlling growth, sight, hearing, smelling, tasting, touching**
19. List the four major blood types. **A, B, AB, O**
20. Name two diseases caused by poor nutrition. **Scurvy, rickets, goiter, anemia**
21. Name the three kinds of respiration. **External, internal, cellular**

22. Name two purposes of melanin. **Protecting skin from ultraviolet light; protecting eyes from ultraviolet light; giving color to eyes, skin, and hair**

23. Based on what we learn from the Bible, God created all of creation as well as Adam and Eve without sickness or death. Why do we see diseases and death in the world today? **Because of the Fall of man, our bodies no longer work as well as they used to, and are now subject to disease and death; because of the curse which was the punishment for man's sin.**

24. Scientists are amazed at how complex the human body is, and they learn new things about it every day. How does this complexity point to a Creator? **Answers will vary. The student should point out that the enormous complexity and organization seen in the human body cannot be a product of chance or the process of naturalistic evolution.**

The World of Animals — Final Exam Answer Key
Lessons 1–34

Match each animal group with its unique characteristic.

1. **D** Mammals
2. **G** Birds
3. **A** Fish
4. **F** Reptiles
5. **H** Amphibians
6. **B** Arthropods
7. **C** Mollusks
8. **I** Echinoderms
9. **J** Cnidarians
10. **E** Protists

Define the following terms.

11. Invertebrate: **Animal without a backbone**
12. Vertebrate: **Animal with a backbone**
13. Cold-blooded animal: **Cold-blooded animals cannot regulate their body temperature; it is the same as the surrounding temperature.**
14. Warm-blooded animal: **Warm-blooded animals regulate their body temperature to keep it the same regardless of the surrounding temperature.**
15. Moneran: **Monerans are bacteria. They are single-celled organisms without a nucleus.**

Describe how a bird's feet are suited for each task listed below.

16. Swimming in a lake: **Webbed feet for paddling**
17. Perching in a tree: **Three toes facing forward, one toe facing backward for grasping tree branches**
18. Hunting prey: **Sharp claws (or talons) for grasping prey**

Short answer:

19. Describe how a bird is specially designed for flight. **Birds have hollow bones, airfoil shaped wings, contour feathers that point toward the back of the body, special flight feathers, a tail that works like a rudder, and very efficient respiratory and circulatory systems.**
20. Name the three body parts of an insect. **Head, thorax, abdomen**
21. Name the two body parts of a spider. **Cephalothorax, abdomen**

Mark each statement as either True or False.

22. **T** Snakes have a special organ for sensing smell.
23. **T** Cold-blooded animals do not need to eat as often as warm-blooded animals.
24. **F** Turtles can safely be removed from their shells.
25. **T** Cartilaginous fish do not have any bones.
26. **T** Centipedes are arthropods.
27. **F** All crustaceans live in the water.
28. **T** Insects are the most common arthropod.
29. **T** Some creatures can live closely with jellyfish.
30. **F** The best way to kill a starfish is to cut it in half.

Challenge questions

Match the term with its definition.

1. **B** Unguligrade
2. **D** Digitigrade
3. **E** Plantigrade
4. **A** Rumen
5. **G** Abomasum
6. **C** Reticulum
7. **F** Omasum

Mark each statement as either True or False.

8. **F** Darwin's finches prove evolution.
9. **T** Birds have very efficient respiratory systems.
10. **T** Fish sense food by smelling the water.
11. **F** Frogs often confuse one species' call for another.
12. **T** The largest dinosaurs were the sauropods.
13. **T** Marine iguanas live only in the Galapagos Islands.
14. **F** The plastron is the top of a turtle's shell.
15. **T** A turtle's shell is made of the same material as fingernails.

Short answer:

16. Exoskeletons are made from _**chitin/starch**_.

17. The legs of an insect are attached to the _**thorax/middle**_ section of its body.

18. Bioluminescence causes an animal to _**glow**_.

19. Tarantulas have barbed _**hairs**_ that they kick at their enemies.

20. A _**nautilus, octopus, squid, cephalopod**_ moves through the water using jet propulsion.

21. Biomimetics is the study of animals to apply designs to _**human technology**_.

22. Tube worms live near _**hydrothermal vents**_.

23. Plasmodium causes the disease _**malaria**_.

24. _**Antibiotics**_ are used to treat bacterial infections.

25. The stomach of a _**mosquito**_ is needed to complete the sexual reproduction of the plasmodium sporozoan.

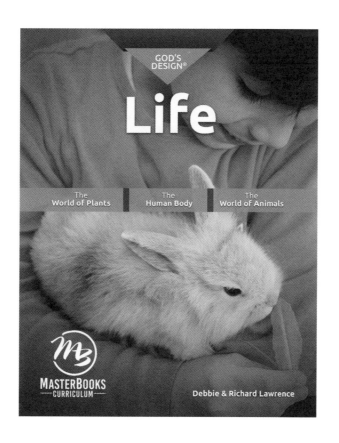

Appendices

for Use with

***God's Design: Life* Series**

The World of Plants Master Supply List

The following table lists all the supplies used for *God's Design for Life: World of Plants* activities. You will need to look up the individual lessons in the student book to obtain the specific details for the individual activities (such as quantity, color, etc.). The letter *c* denotes that the lesson number refers to the challenge activity. Common supplies such as colored pencils, construction paper, markers, scissors, tape, etc., are not listed.

Supplies needed (see lessons for details)	Lesson
☐ Aloe plant	17c
☐ Bread slices (homemade or with no preservatives)	34
☐ Cactus plant	29
☐ Cardboard boxes or shoe boxes	4, 16, 26
☐ Coffee filters	19c
☐ Coffee stirrer	27
☐ Corn meal or yellow sand	21, 31
☐ Craft sticks	6c
☐ Dried moss (from craft store)	32
☐ Encyclopedia (plant and animal)	3
☐ Fern frond	31
☐ Field guide (flowers)	5
☐ Field guide (plants)	27
☐ Field guide (trees)	20
☐ Fingernail polish remover	19c
☐ Flower (composite, such as daisy, sunflower, or zinnia—fresh)	23c
☐ Flower (such as lily—fresh)	23
☐ Flower bulbs (tulips, daffodils, etc.—optional)	12
☐ Food coloring	13, 18
☐ Fruits, nuts, and vegetables	4c, 12, 10, 11, 13, 24, 30, 34c

Supplies needed (see lessons for details)	Lesson
☐ Gelatin mix (yellow)	4
☐ Grass plant	6
☐ Grapes (red and green)	4
☐ Hairspray (aerosol)	34c
☐ Index cards or sketch pad	7, 17, 20, 34c
☐ Jars (1 must have a lid)	8, 9, 30
☐ Knife or scalpel (very sharp)	9, 18, 23, 24
☐ Leaves (fresh)	19c, 20
☐ Magnifying glass (use as needed)	6, 9, 11, 20, 22c, 29, 32
☐ Microscope and slides	4c, 22c, 33c
☐ Modeling clay	21
☐ Peat moss	32c
☐ Photo album with magnetic pages or 3-ring binder	20
☐ Pinecones (scales tightly shut)	10
☐ Pipe cleaners	21
☐ Plants (fast-growing; e.g., mint plants)	16, 28
☐ Plastic cups (clear)	13, 32c
☐ Plastic zipper bags	4, 20, 34
☐ Pollen	22c
☐ Pond water	33c
☐ Poster board/tagboard	2, 12c, 19, 35
☐ Potting soil	6c, 13, 30
☐ Seeds (bean, corn, grass, coconut, radish)	6c, 8, 9, 10c, 11c, 35
☐ Steel wool	8
☐ Straws (flexible)	21, 27
☐ Vinegar	6c
☐ Wooden pencils	7c
☐ Yard stick/meter stick	9

The World of Plants Resource Guide

Suggested Books

Trees of North America by C. Frank Brockman—Great guide for tree identification

Reader's Digest North American Wildlife—A great resource to have for any field trip

The Science of Plants by Jonathan Bocknek—Overview of plants, activities, puzzles for all ages

Plants by Janice VanCleave—Lots of fun activities; science fair ideas

Biology for Every Kid by Janice VanCleave—More fun activities

The Nature and Science of Seeds by Jane Burton and Kim Taylor—Great pictures, overview of less common seeds

Photosynthesis by Alvin Silverstein—In-depth look at the process for upper elementary

How Leaves Change by Sylvia A. Johnson—Great pictures of fall colors, more in-depth information for upper elementary students

God Created the Plants and Trees of the World by Earl & Bonita Snellenberger—Coloring and sticker book for children that teaches about plants and trees from a biblical perspective

Suggested Videos

Newton's Workshop by Moody Institute—Excellent Christian science series; several titles to choose from

Field Trip Ideas

- Creation Museum in Petersburg, KY
- Botanical gardens
- Tour a local nursery, greenhouse, or florist
- Arboretum
- Take a nature walk

Creation Science Resources

Answers Book for Kids Eight volumes by Ken Ham with Cindy Malott—Answers children's frequently asked questions

The New Answers Books 1–4 by Ken Ham and others—Answers frequently asked questions

The Amazing Story of Creation by Duane T. Gish—Scientific evidence for the creation story

Creation Science by Felice Gerwitz and Jill Whitlock—Unit study focusing on creation

Creation: Facts of Life by Gary Parker—In-depth comparison of the evidence for creation and evolution

The Ecology Book by Tom Hennigan & Jean Lightner unveils the intricate nature of God's world and the harmony that was broken by sin.

The World Around You by Gary Parker opens a window to the spectacular environments found on our planet, from deserts to the tropics, with respect to creationism.

The World of Biology by John Tiner is a fascinating look at life - from the smallest proteins and spores, to the complex life systems of humans and animals.

The World of Plants Works Cited

Adams, A. B. *Eternal Quest: The Story of the Great Naturalists*. New York: G.P. Putnam's Sons, 1969.

Bocknek, Jonathan. *The Science of Plants*. Milwaukee: Gareth Stevens Publishing, 1999.

Brockman, C. Frank. *Trees of North America*. New York: Golden Press, 1968.

Burnie, David. *Eyewitness Books Plants*. New York: Alfred A. Knopf, 1989.

Burton, Jane, and Kim Taylor. *The Nature & Science of Seeds*. Milwaukee: Gareth Stevens Publishing, 1999.

"Cobra Lily." http://www..bugbitingplants.com/cobra_lily.php.

Challand, Helwn J. *Plants Without Seeds*. Chicago: Children's Press, 1986.

"A Concise History of the Rose—The King of Flowers." http://www.indiainternational.com. (URL not working)

Croll, Carolyn. *Redoute The Man Who Painted Flowers*. New York: G. P. Putnam's Sons, 1996.

'Espinasse, M. *Robert Hooke*. Berkeley: University of California Press, 1962.

Evans, Erv and Blazich, Frank. *Overcoming Seed Dormancy: Trees and Shrubs*. http://content.ces.ncsu.edu/overcoming-seed-dormancy-trees-and-shrubs.

"The Faces of Science: African Americans in the Sciences." http://webfiles.uci.edu/mcbrown/display/faces.html.

Giannotti, Heike. *The History of Rose Culture in France*. University of Illinois, 2001.

Giesecke, Ernestine. *Outside My Window Flowers*. Des Plaines: Heinemann Library, 1999.

Gish, Duane T. *The Amazing Story of Creation from Science and the Bible*. El Cajon: Institute for Creation Research, 1990.

Gish, Duane T. *Dinosaurs by Design*. Colorado Springs: Creation Life Publishers, 1992.

Greenaway, Theresa. *Mosses & Liverworts*. Austin: Steck-Vaughn Co., 1992.

Greenaway, Theresa. *The Plant Kingdom*. Austin: Steck-Vaughn Co., 2000.

Ham, Ken et al. *The Answers Book*. Green Forest, AR: Master Books, 1992.

Jebens, Brandon. *The Biogeography of Sequoia Sempervirens*. San Fancisco: San Francisco State University, 1999.

Johnson, Sylvia A. *How Leaves Change*. Minneapolis: Lerner Publications Co., 1986.

Karaler, Lucy. *Green Magic Algae Rediscovered*. New York: Thomas Y. Crowell, 1983.

Koerner, L. *Linnaeus: Nature and Nation*. Cambridge: Harvard University Press, 1999.

Lindroth, S. *The Two Faces of Linnaeus*. Berkeley: University of California Press, 1983.

Moore, J. A. *Science as a Way of Knowing*. Cambridge: Harvard University Press, 1993.

Morris, John D., Ph.D. *The Young Earth*. Green Forest, AR: Master Books, 1994.

Parker, Gregory, and others. *Biology: God's Living Creation*. 2nd ed. Pensacola: A Beka Book, 1986.

Omahen, Sharen, *Seed Storage Facility is Modern Day Noah's Ark*. Environmental Report 2004, published by the Office of Environmental Sciences, University of Georgia College of Agricultural and Environmental Sciences, pg. 15; https://athenaeum.libs.uga.edu/bitstream/handle/10724/33818/EnvrionReport2004.pdf?sequence=1

"Pierre-Joseph Redoute." http://www.globalgallery.com/knowledgecenter/artist_biography/pierre+joseph+redoute.

Reader's Digest. *North American Wildlife*. Pleasantville: Reader's Digest Association, 1982.

Rogers, Kirsteen et al. *Usborne Science Encyclopedia*. London: Usborne Publishing, 2002.

Rudwick, M. J. S. *The Meaning of Fossils*. Chicago: University of Chicago Press, 1985.

Schiebinger, L. "The Loves of the Plants." *Scientific American*. February 1996: 110-115.

Selsam, Millicent E. *Mushrooms*. New York: William Morrow & Co., 1986.

Silverstein, Alvin, and others. *Photosynthesis*. Brookfield: Twenty-first Century Books, 1998.

Steele, DeWitt. *Investigating God's World*. Pensacola: A Beka Book Publications, 1986.

"This Person in Black History 4: George Washington Carver." http://sps.k12.mo.us/historyday/feb/carver.htm.

"The Truly Amazing Redwood Tree." http://www.treesofmystery.net/redwood-trees.php.

VanCleave, Janice. *Biology for Every Kid*. New York: John Wiley & Sons, Inc., 1990.

VanCleave, Janice. *Plants*. New York: John Wiley & Sons, Inc., 1997.

Wexler, Jerome. *From Spore to Spore*. New York: Dodd, Mead & Co., 1985.

The Human Body Master Supply List

The following table lists all the supplies used for *God's Design for Life: The Human Body* activities. You will need to look up the individual lessons in the student book to obtain the specific details for the individual activities (such as quantity, color, etc.). The letter *c* denotes that the lesson number refers to the challenge activity. Common supplies such as colored pencils, construction paper, markers, scissors, tape, etc., are not listed.

Supplies needed (see lessons for details)	Lesson
☐ Aluminum foil	20
☐ Anatomy book	5–34
☐ Balloons	28
☐ Bible	1, 35
☐ Candy sprinkles	26
☐ Chicken bones	6c
☐ Cinnamon, peppermint, and other spices	18
☐ Dental floss	21
☐ Dissection kit	25c
☐ DNA model kit (optional)	33c
☐ Eyedropper	31c
☐ Food coloring	31c
☐ Fruits, nuts, and vegetables	18
☐ Gel pens (washable)	5
☐ Gloves (rubber or latex)	25c
☐ Hand lotion	29
☐ Heart (from a cow or sheep)	25c
☐ Index cards	13, 31
☐ Jelly beans (white)	26
☐ Knife or scalpel (very sharp)	25c
☐ Lemon juice	18

Supplies needed (see lessons for details)	Lesson
☐ Light corn syrup	26, 31c
☐ Magnifying glass	8c
☐ Mirror	1, 21, 29
☐ Modeling clay	12, 20
☐ Newsprint (or other large roll of paper)	34
☐ Paper towels	18
☐ Paper fasteners (brads)	2, 4
☐ Plaster of Paris	20
☐ Plastic zipper bags	15
☐ Poster board/tagboard	20
☐ Rubber/plastic gloves	25c
☐ Red Hots candies	26
☐ Rubber bands	7c
☐ Ruler	11
☐ Salt	18
☐ Steak (or other meat—raw)	8c
☐ Stopwatch	9, 11, 24, 28
☐ Straight pins	15
☐ Sugar	18
☐ Tacks	7c
☐ Tape measure (cloth—the kind used for sewing)	28
☐ Tape measure (metal)	4c
☐ Toothbrush	21
☐ Toothpaste	21
☐ Toothpicks	14
☐ Vinegar	6c, 18
☐ Wooden pencils	7c
☐ Yard stick/meter stick	9, 31c

The Human Body Resource Guide

Suggested Books

Human Anatomy in Full Color from Dover Publications—Highly recommended

Human Anatomy Coloring Book from Dover—Detailed coloring pages

Biology for Every Kid by Janice VanCleave—Many fun activities

Understanding Your Senses by Rebecca Treays—Good pictures and explanations

Body by Design by Dr. Alan Gillen—Basic anatomy and physiology from a creationist viewpoint

God Created the People of the World by Earl and Bonita Snellenberger—An information-packed coloring and sticker book with the biblical history of people and people groups

Suggested Videos

Newton's Workshop by Moody Institute—Excellent Christian science series

The Hearing Ear and the Seeing Eye by Dr. David Menton—Amazing features God created

Fearfully and Wonderfully Made by Dr. David Menton—Discusses the miracle of birth

Field Trip Ideas

- Creation Museum in Petersburg, Kentucky
- Hospital
- Science museum with human body exhibits
- Doctor's Office

Other Ideas

Build a model of the human body.

Dissect a sheep's or cow's heart, brain, or eye. Examining animal parts can give a more realistic idea of what is inside the human body.

Creation Science Resources

Answers Book for Kids Eight volumes by Ken Ham with Cindy Malott—Answers children's frequently asked questions

The New Answers Books 1–4 by Ken Ham and others—Answers frequently asked questions

Wonders of the Human Body series by Dr. Tommy Mitchell builds your faith as you discover the amazing handiwork of their Creator in the human body's incredible design from cells to the organs themselves.

Body by Design by Alan Gillen presents the basic anatomy and physiology in each of 11 body systems from a creational viewpoint while exposing faulty evolutionistic reasoning.

The Human Body Works Cited

"A ,B, Cs of Brain Tumors." http://www.brain-surgery.com/primer.html.

Adams, A. B. *Eternal Quest: The Story of the Great Naturalists*. New York: G.P. Putnam's Sons, 1969.

Collins, David R. *God's Servant at the Battlefield Florence Nightingale*. Milford: Mott Media, 1985.

'Espinasse, M. *Robert Hooke*. Berkeley: University of California Press, 1962.

"Florence Nightingale." http:///www.countryjoe.com/nightingale/.

Gish, Duane T., Ph.D. *The Amazing Story of Creation From Science and the Bible*. El Cajon: Institute for Creation Research, 1990.

Gish, Duane T., Ph.D. *Dinosaurs by Design*. Colorado Springs: Creation-Life Publishers, 1992.

"Gregor Mendel." http://www.accessexcellence.org.

Ham, Ken et al. *The Answers Book*. Green Forest, AR: Master Books, 1992.

Harcup, John W. *Human Anatomy in Full Color*. Mineola: Dover Publications, Inc., 1996.

"How the Immune System Works." http://health.howstuffworks.com/human-body/systems/immune/immune-system.htm.

Jefferies, David. *The Human Body*. Huntington Beach: Teacher Created Materials, Inc., 1993.

Koerner, L. *Linnaeus: Nature and Nation*. Cambridge: Harvard University Press, 1999.

Lindroth, S. *The Two Faces of Linnaeus*. Berkeley: University of California Press, 1983.

Mangiardi, John R., and Howard Kane. *History of Brain Surgery*. www.brain-surgery.com.

"Mapping the Motor Cortex." http://www.pbs.org/wgbh/aso/tryit/brain/cortexhistory.html.

Moore, J. A. *Science as a Way of Knowing*. Cambridge: Harvard University Press, 1993.

Morris, John D. *The Young Earth*. Green Forest, AR: Master Books, 1994.

Parker, Gregory, and others. *Biology God's Living Creation*. Pensacola: A Beka Books, 1997.

Roehm, Michelle. *Girls Who Rocked the World*. Hillsboro: Beyond Words Publishing, 2000.

Rudwick, M.J. S. *The Meaning of Fossils*. Chicago: University of Chicago Press, 1985.

Schiebinger, L. "The Loves of the Plants." *Scientific American*. February 1996: 110–115.

Steele, DeWitt. *Investigating God's World*. Pensacola: A Beka Book, 1986.

Treays, Rebecca. *Understanding Your Senses*. London: Usborne Publishing, Ltd., 1997.

VanCleave, Janice. *Biology for Every Kid*. New York: John Wiley & Sons, 1990.

VanCleave, Janice. *Plants*. New York: John Wiley & Sons, 1990.

VanderMeer, Ron, and Ad Dudnik. *The Brain Pack*. Datchet: VanderMeer Publishing, 1996.

The World of Animals Master Supply List

The following table lists all the supplies used for *God's Design for Life: World of Animals* activities. You will need to look up the individual lessons in the student book to obtain the specific details for the individual activities (such as quantity, color, etc.). The letter *c* denotes that the lesson number refers to the challenge activity. Common supplies such as colored pencils, construction paper, markers, scissors, tape, etc. are not listed.

Supplies needed (see lessons for details)	Lesson
☐ 3-ring binder	2
☐ Ammonia	28
☐ Baking dish	28
☐ Balloons	21c, 27c
☐ Bible	35
☐ Bird feeder (optional)	8
☐ Butterfly larvae (caterpillars)	23
☐ Chocolate chips (mini size)	29
☐ Crushed chocolate cookies	31
☐ Dissection kit	29c
☐ Dividers with tabs (12 or 13 per student)	2
☐ Encyclopedia (animal)	all
☐ Face paint	18
☐ Fake fur or felt	7
☐ Feather (can purchase at craft store)	9
☐ Field guide (birds)	8
☐ Field guide (sea shells)	27
☐ Flexible wire	24
☐ Flour	21c
☐ Food coloring	28
☐ Goldfish snack crackers	11
☐ Gummy worms	31
☐ Hair/fur from 2 or more mammals	3
☐ Hydras (live)	28c
☐ Index cards	22
☐ Instant chocolate pudding	31

Supplies needed (see lessons for details)	Lesson
☐ Liquid bluing	28
☐ Magnifying glass	9, 24, 25
☐ Marshmallows (large and small)	24
☐ Microscope and slides (optional)	32
☐ Modeling clay	13, 25, 26c
☐ Newspaper	21c
☐ Owl pellet (optional)	10
☐ Pipe cleaners	22, 24, 26c
☐ Plastic zipper bags	7
☐ Pond water (optional)	32
☐ Poster board/tagboard	7, 29
☐ Rubber/plastic gloves	29c
☐ Salt	28
☐ Sand dollar (dead and dried; check at craft store—optional)	29
☐ Spider web (optional)	24
☐ Starfish (dead and dried; check at craft store—optional)	29
☐ Salt dough	29
☐ Sequins or flat beads	16
☐ Sea shells	27
☐ Soap (anti-bacterial hand)	33
☐ Sponge (natural—optional)	30
☐ Sponge (synthetic)	30
☐ Starfish (preserved for dissection)	29c
☐ Stopwatch	6
☐ String	21c
☐ Styrofoam balls	22
☐ Tadpoles and tank (optional)	15
☐ Tape (cloth)	19
☐ Tempera paints	30
☐ Toothbrush	6
☐ Toothpicks	22, 24
☐ Yarn	32

The World of Animals Resource Guide

Suggested Books

Breathtaking Birds by Buddy and Kay Davis—Animal encyclopedia from a creation perspective

Magnificent Mammals by Buddy and Kay Davis—Animal encyclopedia from a creation perspective

Sensational Sea Creatures by Buddy and Kay Davis—Animal encyclopedia from a creation perspective

Reader's Digest North American Wildlife—A great resource to have for any field trip

Mammals Scholastic Voyages of Discovery by Scholastic Books—Interactive fun

Birds by Carolyn Boulton—Lots of suggested activities, good pictures

Play and Find Out About Bugs by Janice VanCleave—Great experiments

Jellyfish by Leighton Taylor—Good explanation of lifecycle, great pictures

Zoo Guide by Answers in Genesis—A guide to over 100 animals from a creationist perspective

Aquarium Guide by Answers in Genesis—Take it to the aquarium with you

Museum Guide by Answers in Genesis—Take it to the natural history museum with you

God Created Series by Earl and Bonita Snellenberger—Coloring and sticker books that present God's creation to young children

Suggested Videos

Newton's Workshop by Moody Institute—Excellent Christian science series

Incredible Creatures that Defy Evolution Three volumes by Exploration Films—Learn about amazing design features

Exploring the Wildlife Kingdom by Exploration Films—Evolution-free nature videos

Life's Story by NPN Videos—Shows how the interactions among living things could not have happened through evolution

We highly recommend purchasing one or more of the following to supplement the activities in this book:
Owl pellets
Frog habitat
Butterfly habitat
Dissection supplies

Field Trip Ideas

- Creation Museum and Ark Encounter in Petersburg, KY
- Farm or dairy
- Zoo, aquarium, or butterfly museum
- Fish hatchery
- Wildlife area

Creation Science Resources

Answers Book for Kids Eight volumes by Ken Ham with Cindy Malott—Answers children's frequently asked questions

The New Answers Books 1–4 by Ken Ham and others—Answers frequently asked questions

Dinosaurs of Eden by Ken Ham which takes you on a breathtaking trip across time to the biblical foundation of dinosaurs.

Dinosaurs: Marvels of God's Design by Dr. Tim Clarey is a thoroughly researched, definitive guide to dinosaurs for Christian readers!

Creation Science by Felice Gerwitz and Jill Whitlock—Unit study focusing on creation

Creation: Facts of Life by Gary Parker—In-depth comparison of the evidence for creation and evolution

Dinosaurs for Kids by Ken Ham—Learn the true history of dinosaurs

The World of Animals Works Cited

Adams, A. B. *Eternal Quest: The Story of the Great Naturalists*. New York: G.P. Putnam's Sons, 1969.

Bargar, Sherie, and Linda Johnson. *Rattlesnakes*. Vero Beach: Rourke Enterprises, Inc., 1986.

"Birds." http://www.earthlife.net/birds.

Cardwardine, Mark, et.al. *Whales, Dolphins & Porpoises*. Sydney: US Weldon Owen Inc., 1998.

Chinery, Michael. *Butterfly*. Mahwah: Troll Associates, 1991.

Chinery, Michael. *Shark*. Mahwah: Troll Associates, 1991.

Coldrey, Jennifer. *Shells*. New York: Dorling Kindersley, Inc., 1993.

Cole, Joanna. *A Bird's Body*. New York: William Morrow & Co., 1982.

Cousteau Society. *Corals: The Sea's Great Builders*. New York: Simon & Shuster, 1992.

'Espinasse, M. *Robert Hooke*. Berkeley: University of California, 1962.

Evans, J. Edward. *Charles Darwin: Revolutionary Biologist*. Minneapolis: Lerner Publications, 1993.

Fleisher, Paul. *Gorillas*. New York: Benchmark Books, 2001.

Gish, Duane T., Ph.D. *The Amazing Story of Creation*. El Cajon: Institute for Creation Research, 1990.

"Georges Cuvier." http://www.ucmp.berkeley.edu/history/cuvier.html.

Gowell, Elizabeth Tayntor. *Whales and Dolphins: What They Have in Common*. New York: Franklin Watts, 1999.

Ham, Ken. *The Great Dinosaur Mystery Solved!* Green Forest: Master Books, 1999.

Hird, Ed. "Dr. Louis Pasteur: Servant of All." *Deep Cove Crier*. December 1997.

Jackson, Tom. *Nature's Children Rattlesnakes*. Danbury: Grolier Educational, 2001.

Kalman, Bobbie, and Allison Larin. *What Is a Fish?* New York: Crabtree Publishers, 1999.

Koerner, L. *Linnaeus: Nature and Nation*. Cambridge: Harvard University, 1999.

Lacey, Elizabeth A. *The Complete Frog: A Guide for the Very Young Naturalist*. New York: Lothrop, Lee & Shepard Books, 1989.

Landau, Elaine. *Sea Horses*. New York: Children's Press, 1999.

Lindroth, S. *The Two Faces of Linnaeus*. Berkeley: University of California Press, 1983.

"Louis Pasteur." http://web.ukonline.co.uk/b.gardner/pasteur.htm. (URL not working)

Maynard, Thane. *Primates Apes, Monkeys, Prosimians*. New York: Franklin Watts, 1994.

Markle, Sandra. *Outside and Inside Kangaroos*. New York: Atheneum Books for Young Readers, 1999.

Moore, J. A. *Science as a Way of Knowing*. Cambridge: Harvard University Press, 1993.

Morris, John D., Ph.D. *The Young Earth*. Green Forest, AR: Master Books, 1994.

National Geographic Book of Mammals. Washington, D.C.: National Geographic Society, 1998.

Parker, Gregory et al. *Biology: God's Living Creation*. Pensacola: A Beka Book, 1997.

Parker, Steve. *Charles Darwin and Evolution*. London: HarperCollins Publishers, 1992.

Ross, Michael E. *Wormology*. Minneapolis: Carolrhoda Books, Inc., 1996.

Rudwick, M.J. S. *The Meaning of Fossils*. Chicago: University of Chicago Press, 1985.

Scholastic Voyages of Discovery Mammals. New York: Scholastic, Inc., 1997.

Stone, Lynn M. *Tasmanian Devil*. Vero Beach: Rourke Corporation, Inc., 1990.

Swan, Erin Pembrey. *Meat-eating Marsupials*. New York: Franklin Watts, 2002.

Swan, Erin Pembrey. *Primates: From Howler Monkeys to Humans*. New York: Franklin Watts, 1998.

Taylor, Leighton. *Jellyfish*. Minneapolis: Lerner Publishing Co., 1998.

VanCleave, Janice. *Biology for Every Kid*. New York: John Wiley & Sons, Inc., 1990.

VanCleave, Janice. *Insects and Spiders*. New York: John Wiley & Sons, Inc., 1998.

VanCleave, Janice. *Play and Find Out About Bugs*. New York: John Wiley & Sons, Inc., 1999.

Walters, Martin. *The Simon & Schuster Young Readers' Book of Animals*. New York: Simon & Schuster, Inc., 1990.

The World of Plants — Glossary

Adventitious roots Roots that grow in unexpected places or in unexpected ways
Aerial roots Roots that take water from the air
Aggregate fruit Formed from one flower with multiple pistils and ovaries
Algae Plantlike organisms that often live in the water; Kingdom Protista
Alternate leaf arrangement One leaf grows from each node on alternating sides of the stem
Anatomy The study of the human body
Angiosperm Plant that reproduces with flowers, fruit, and seeds
Annual Plant that completes lifecycle in one growing season
Anther Part of the stamen that produces pollen
Axillary bud/Lateral bud Bud growing from side of stem

Bark Dead, hardened epidermis cells in woody stems
Biennial Plant that completes lifecycle in two growing seasons
Binomial classification Two-name system of classification developed by Carl Linnaeus
Botany The study of plants

Cambium cells Cells that divide to produce more xylem and phloem
Carnivorous Meat eating
Catalyst Substance that speeds up a chemical reaction
Cell Smallest unit of an organism that can survive on its own
Cell membrane Outer coating or "skin" of a cell
Cell wall Rigid outermost layer of a plant cell
Cereal grass Grains such as wheat and oats
Chlorophyll Green substance in chloroplasts that makes photosynthesis possible
Chloroplast Part of a cell that transforms sunlight into food (glucose)
Conifer Plant that reproduces with seeds in cones
Cotyledon Food stored in seed to supply nourishment to new plant
Cross-pollination Flower is pollinated with pollen from another plant
Cuticle Top layer of epidermis
Cytoplasm Liquid that fills a cell

Deciduous Trees that lose their leaves in the winter
Diatoms Yellow algae with silica in their cell walls
Dicot Seed with two cotyledons
Disperse/Dispersal Movement of seed away from parent plant
Dormant A condition in which the seed is inactive or "asleep"

Embryo "Baby" plant inside a seed
Endosperm Additional nutrients absorbed by cotyledon during germination of monocots
Epidermis Outer layer of cells in a young stem
Evergreen Trees that do not lose their leaves in the winter

Fibrous roots Roots spread out in many directions
Filament Stalk of the stamen that supports the anther
Flowers Organs that produce fruits and seeds for reproduction
Forage grass Taller grass eaten by grazing animals
Fronds "Leaves" of a fern
Fruit Ripened ovary
Fungi Organisms that cannot make their own food including mushrooms and yeast; Kingdom Fungi

Geotropism The ability to sense up and down, response to gravity
Germinate When seeds begin to grow
Guard cells Cells which open and close the stomata
Gymnosperm Plant that reproduces with cones and seeds

Haustoria Shoots sent from parasitic roots to tap into another plant's roots
Heartwood Dead xylem cells in center of tree that no longer transport materials
Heliotropism/Phototropism The ability to sense light; response to light
Herbaceous plants Plants with bendable stems
Hilum Location on seed where it was attached to the ovary of the plant
Host Plant from which a parasite takes nutrients
Hydrotropism The ability to sense water, response to water

Internode Stem between two nodes

Kingdom Group of living things that have broad common characteristics

Leaves Organs that manufacture food for the plant

Mitochondria Part of a cell that breaks down food into energy
Monocot Seed with one cotyledon
Multiple fruit Formed when several flowers form fruit that fuse together

Nectar Sweet liquid that attracts pollinators
Node Point where leaf attaches to stem
Nucleus Control center or "brain" of a cell

Opposite leaf arrangement Two leaves grow on opposite sides of the stem from one node
Organ A group of tissues working together to perform a function
Ornamental grass Very tall grass used for landscaping
Ovary Part of the pistil that produces the ovules
Ovule Unfertilized seed, egg

Palmate Palm-like venation
Parasite Plant that gains nutrients by tapping into and taking them from other plants
Parasitic roots Roots that tap into another plant's roots to steal nutrients and water
Passenger plant Attached to other plants but does not harm them
Perennial Plant that grows year after year
Petal Part of the flower that attracts pollinators—often brightly colored and scented
Petiole The part of the leaf that attaches to the stem
Phloem Tubes that transport food from leaves back down to the roots
Photosynthesis Process that changes light, water, and carbon dioxide into sugar and oxygen
Phylum, class, order, family, genus, species Different levels of how living things in a kingdom are divided into groups by common characteristics
Pinnate Feather-like venation
Pistil Female part of the flower—contains ovules
Pistillate Flower that produces only pistils
Plumule Part of embryo which develops into the stem and leaves
Pneumatophore Roots that grow above ground to absorb oxygen from the air
Pollen Fine powder needed for reproduction

Pollination Uniting of pollen with an ovule
Pollinator Animal that distributes pollen
Primary consumers Animals that eat plants
Prop roots Roots growing out from the side of a stem then into the ground to provide stability

Radicle Part of the embryo which develops into the roots
Respiration Exchange of oxygen and carbon dioxide in living cells
Rhizomes Special underground stems that grow horizontally
Root cap Covering that protects tip of root
Roots Organs that anchor plants and absorb water and nutrients
Rosette leaf arrangement Leaves grow from the bottom of the stem

Sapwood Area of stem with active xylem and phloem cells
Secondary consumers Animals that eat the primary consumers
Seed coat Protective covering on outside of seed
Self-pollination Flower is pollinated with pollen from the same plant
Sepal Part of the flower that protects the developing flower
Shoot New stem growth
Simple fruit Formed from one flower with one pistil and one ovary
Spores Reproductive organs of non-flowering plants
Stamen Male part of the flower—produces pollen
Staminate Flower that produces only stamens
Stems Organs that hold up plants and provide their basic shape
Stigma Part of the pistil that receives the pollen
Stolons/Runners Special stems that produce new plants
Stomata Holes on the underside of a leaf
Style Stalk of the pistil that supports the stigma
Succulents Plants that have the ability to store large amounts of water

Taproot One large central root with many smaller roots branching out
Taxonomy Method of classifying living things
Tendrils Special stems that grab onto things
Terminal bud Bud at the end of a stem
Thorns Special stems for protection

Tissue Group of cells working together to perform a function
Tropism Plant response to a particular stimulus/condition
Tubers and bulbs Special stems that store food underground
Turf grass Short grass used for lawns

Vacuole Food storage location in a cell
Vascular tissue Series of tubes similar to blood vessels for transporting nutrients and other chemicals throughout a plant
Vegetative reproduction/Vegetative propagation Reproduction using part of the plant instead of seeds to start a new plant

Venation Arrangement of a leaf's veins
Vine Plants that grow on other plants or structures for support but have their own root systems

Whorled leaf arrangement Three or more leaves grow from one node around a stem
Woody plants Plants with stiff woody stems

Xylem Tubes that transport water and nutrients from roots to the rest of the plant

Zoology The study of animals

The World of Plants — Challenge Glossary

Abiogenesis/Chemical evolution Idea that at one time simple life came from nonliving chemicals; a modern version of spontaneous generation
Algin Sticky substance found in kelp, a brown algae
Anaphase Third phase of mitosis in which the chromosomes are pulled apart
Anthocyanin Red or purple pigment in plants

Berry Succulent fruit with multiple seeds
Bract Bright colored leaf for attracting pollinators

Capillarity Movement of water due to attraction of water molecules for each other
Carotene Yellowish-orange pigment in plants
Carrageenin Gelatin-like substance from red algae
Chemotropism Response to chemicals
Cloning Offspring have identical DNA to parent
Composite flowers Collection of hundreds of tiny flowers on one stalk
Compound leaf Several leaflets off of a single petiole
Crown Branches of a tree
Cutting Propagation by cutting a stem and stimulating it to grow new roots
Cytokinesis The division of the cytoplasm

Deliquescent branching Strong growth in lateral buds resulting in horizontal growth habit
Dichotomous key Chart presenting two options at each level for classification

Diffusion Movement of molecules from an area of higher concentration to an area of lower concentration
Disk flowers Flowers comprising the head of a composite flower
Dispersing agent External force aiding in dispersal
Double dormancy Seeds require both scarification and stratification to germinate
Drupe Succulent fruit with a single hard seed
Dry fruit Simple fruit with a dry outer layer

Entire margin Smooth leaf margin
Ephemeral Plant with a very short lifecycle
Epigeal germination Cotyledons move above ground after germination
Epiphyte A plant that grows on another plant using the host only for support
Excurrent branching Strong growth in terminal buds resulting in a vertical growth habit
External dormancy/Seed coat dormancy Dormancy lasts until seed coat is softened and/or broken

Fiddlehead Developing petiole of a fern frond
Filament algae Algae connected together end to end to form long strings

GMO Genetically modified organism
Genetic modification Modifying a plant's genes to obtain desired results
Glucose Sugar produced in photosynthesis

Grafting   Propagation by combining a bud onto a rootstock
Grain   Dry fruit of the grass family
Growth habit   Way the branches of a tree grow; a tree's shape

Head   Center of a composite flower
Hypogeal germination   Cotyledons remain below ground after germination

Internal dormancy/Embryo dormancy   Dormancy lasts until certain temperature or moisture requirements are met

Law of Biogenesis   Life can only come from life
Leaf margin   The edge of a leaf
Legume   Dry fruit with a pod around the seeds
Lobed margin   Deeply indented leaf margin

Meiosis   Cell division that results in reproductive cells
Metaphase   Second phase of mitosis in which the chromosomes line up in the middle of the cell
Mitosis/Fission   Cell division resulting in two daughter cells identical to the original cell

Nectar/Pollen guide   Markings on flowers to direct pollinators to the nectar
Negative tropism   Movement away from the stimulus
Nut   Dry fruit with hard outer shell

Osmosis   Diffusion through a membrane

Peat   Layers of decaying peat moss
Pome   Succulent fruit with a papery core around the seeds
Positive tropism   Movement toward the stimulus
Primary growth   Growth that results in longer roots, stems, etc.
Prophase   First phase of mitosis in which the nuclear envelope dissolves

Ray flowers   Flowers that look like petals on a composite flower
Receptacle   Where the flower attaches to the stem
Root hairs   Tube-like projections on roots that are responsible for absorbing most of the water
Rootstock   Root and stem grown specifically for grafting

Scarification   Actions that result in breaking of seed coat
Scion   Stem or bud that is grafted onto a rootstock
Secondary growth   Growth that results in thicker roots, stems, etc.
Seed dormancy   Seeds will not germinate because certain conditions have not been met
Simple leaf   Only one leaf per petiole
Spines   Needle-like leaves designed to conserve water
Spontaneous generation   Belief that animals were suddenly produced by their surroundings
Starch   A string of glucose molecules linked together
Stratification   Seeds experience an extended period of cold temperature
Succulent fruit   Simple fruit with a thick fleshy outer layer
Succulent leaves   Fleshy leaves that store water
Sucrose   More complex sugar formed by combining two glucose molecules

Telophase   Final phase of mitosis in which nuclear envelopes develop around the chromosomes and cell divides into two separate cells
Thermotropism   Response to changes in temperature
Thigmotropism   Response to touch
Toothed margin   Jagged leaf margin
Transpiration   Evaporation of water from plants

Vascular bundles   Groups or bundles of xylem and phloem

Xanthophyll   Yellow pigment in plants

Zone of cell division   Region where cells are actively dividing
Zone of differentiation/Zone of maturation   Region where cells line up to form vascular tissue
Zone of elongation   Region of root that lengthens due to lengthening of cells

The Human Body — Glossary

Alveoli   Sacs where exchange of gases takes place in the lungs (singular is *alveolus*)
Amplitude   How high waves are; in sound this determines loudness
Antibodies   Proteins that attack and destroy invading substances
Aorta   Artery taking blood from the heart to the body
Arteries   Blood vessels that carry blood away from the heart
Asthma   Swelling of the bronchial tubes
Atrium   Upper chamber in the heart that receives blood (plural is *atria*)
Auditory canal   Opening of the ear

Ball and socket joint   Moves in two planes, rotates in place
Bicep and tricep   Muscle pair in upper arm
Bicuspids   Bumpy teeth for grinding, next to canine teeth
Bone marrow   Material in the center of long bones that produces blood cells
Brain stem   Connection between the rest of the brain and the spinal cord, regulates automatic functions
Bronchi/Bronchial tubes   Tubes entering the lungs

Capillaries   Blood vessels that connect arteries and veins
Carbohydrates   Sugars and starches
Carpals   Bones of the wrist
Cartilage   Smooth, slippery material forming cushion between bones
Cell membrane   Protective shell or "skin" of the cell
Cell   Smallest part of the human body that can function on its own
Central nervous system   Brain and spinal cord
Cerebellum   Lower part of the brain, controlling balance and muscle movement
Cerebral cortex   Outermost layer of the cerebrum
Cerebrum   Upper part of the brain, controlling thought and memory
Cilia   Tiny hairs covering many body tissues
Circulatory system   System of heart, blood, and blood vessels that carry oxygen and nutrients throughout the body
Clavicle   Collarbone
Cochlea   Fluid-filled part of the ear that changes vibrations to electrical signals
Cones   Special cells in the retina to detect color

Cornea   Front of the eye
Coronary arteries   Blood vessels carrying oxygen-rich blood to the heart muscle
Corpus callosum   Large bundle of nerves connecting two sides of the cerebrum
Cranium   Skull
Cuspids/Canine teeth   Pointed teeth for tearing
Cytoplasm   Liquid inside the cell

Dermis   Middle layer of skin containing nerves, sweat glands, and other functions
Diaphragm   Muscle below lungs that expands the chest cavity when it contracts
Digestive system   System of organs that removes nutrients from food you eat
Dominant gene   "Stronger" gene, or predominant characteristic that is passed down from parent to child

Elastin   Elastic-like fibers that allow skin to stretch
Ellipsoid joint   Moves in two planes without rotating
Enzymes   chemicals that help with the digestion process
Epidermis   Top layer of skin
Esophagus   Tube connecting your mouth with your stomach

Femur   Large bone in upper leg
Fibrin   Stringy substance that helps form scabs
Fibula   Smaller bone in lower leg
Flat bones   Scapula, cranium, ribs, and pelvis
Frequency   How close waves are together; in sound this determines pitch
Friction skin   Ridged skin found on feet and hands, for gripping

Genes   Tiny bits of information contained in the body's cells, stored in DNA
Gliding joint   Bones slide over each other
Gluteus maximus   Muscle in rear

Hair follicle   Area in skin where hair is formed
Malleus, incus, and stapes   Tiny bones in the middle ear
Hemoglobin   Chemical in red blood cells that turns bright red in the presence of oxygen
Hinge joint   Moves in only one direction

Hippocampus   Part of the brain that stores and retrieves short-term memory
Humerus   Bone in upper arm

Immune system   System of organs that work together to fight off infection
Incisors   Sharp straight teeth for biting
Inferior vena cava   Vein bringing blood to the heart from the lower body
Integumentary system   System of organs including the skin, nails, and hair
Involuntary muscles   Muscles that are controlled without conscious thought
Iris   Colored part of the eye, controls size of pupil
Irregular bones   Bones that don't fit into any other category for shape

Keratin   Substance that makes skin tough

Large intestine   Tube where water is removed from unusable material
Larynx   Voice box
Lens   Part of the eye that focuses the image
Long bones   Arm and leg bones
Loops, whorls, and arches   Patterns formed by fingerprints
Lymph nodes   Bean-shaped organs in the lymph system containing white blood cells
Lymph system   System of vessels and lymph nodes that help eliminate invading substances
Lymph   Clear liquid in the lymph system that collects germs to be carried to the lymph nodes for elimination

Mandible   Jawbone
Medulla oblongata   Lowest part of the brain stem
Melanin   Pigment cells in the skin
Melanocytes   Cells which produce melanin
Metacarpals   Bones in the hand
Mitochondria   Break down nutrients to provide energy for the cell
Molars   Large bumpy teeth for grinding, in back of mouth
Muscular system   System of muscles that move the bones and other parts of the body

Nervous system   System of nerves that control the body
Neuron   Nerve cell
Nucleus   Brain of the cell

Optic nerve   Nerve that transmits image to the brain

Organ   Many tissues working together to perform a function

Patella   Kneecap
Pectoral   Muscles in upper chest
Peripheral nervous system   Nerves and other sensory organs
Phalanges   Small bones in your fingers and toes
Pharynx   Back of the throat
Pituitary gland   Part of the brain that controls growth
Pivot joint   Rotates only
Plaque   Sticky substance that builds up on teeth
Plasma   Yellowish liquid that transports blood cells
Platelets   Cells that help to close a wound
Pneumonia   Illness causing fluid in the lungs
Pulmonary artery   Artery carrying blood from the heart to the lungs
Pulmonary veins   Veins carrying blood from the lungs to the heart
Pupil   Opening in the eye that allows light to enter

Radius   Bone in the lower arm on the side with your thumb
Red blood cells   Cells that carry oxygen and nutrients to the body
Reflex   Automatic reaction that does not require a signal to go to the brain
Respiratory system   System of organs including the lungs whose main function is to get oxygen to the blood
Retina   Back of the eye that detects the image
Rods   Special cells in the retina to detect differences in light

Saddle joint   Moves in two planes, shaped like a saddle
Scapula   Shoulder blade
Sebaceous gland   Gland in the skin that secretes oil
Short bones   Bones of the fingers, toes, hands, and feet
Skeletal system   System of bones that provides strength and gives the body its general shape and size
Small intestine   Tube where nutrients are absorbed
Smooth muscle   Muscles that line the digestive tract and other internal organs
Spleen   Organ that filters dangerous organisms from the blood
Sternum   Bone in central chest connecting ribs
Stomach   Organ where food is broken down into smaller molecules

Subcutaneous   Lowest layer of skin, connecting skin to the body tissues
Superior vena cava   Vein bringing blood to the heart from the upper body

Tendon   Cord-like structure connecting muscles to bones
Thalamus   Part of the brain that routes messages within the brain
Thymus   Organ producing special white blood cells called T-cells
Tibia   Larger bone in the lower leg
Tissue   Many cells working together to perform a function
Trachea   Tube leading to the lungs
Trapezius   Muscles in upper back

Ulna   Bone in the lower arm on the side away from your thumb

Vacuole   Food storage area in the cell
Veins   Blood vessels that carry blood toward the heart
Ventricle   Lower chamber in the heart that pushes out blood
Vertebrae   Bones in the spine
Villi   Small finger-like projections inside the small intestine that absorb nutrients
Voluntary muscles   Muscles that move when you actively think about movement

White blood cells   Cells that help fight infection

The Human Body — Challenge Glossary

Adenine   Base used in DNA, must be paired with thymine
Albinism   Condition where the body does not produce melanin
Antibiotics   Chemicals that inhibit the growth of bacteria and other microorganisms
Antigens   Identification tags on cells
Appendicular skeleton   Outer bones including arms, legs, hips, shoulders, feet, and hands
Aqueous humor   Liquid between the cornea and the lens
Axial skeleton   Central bones including skull, face, neck, spine, and ribs
Axon   Part of the neuron that carries signals away from the cell body

Base pair   Two bases connected together to form a rung of the DNA molecule
Bile   Chemicals produced by the liver and stored in the gall bladder to be released into the small intestine and used for digestion
Blood pressure   Pressure placed on the walls of the blood vessels
Blood type   Determined by the absence or presence of certain antigens in the blood
Braille   System of raised dots to represent letters and other symbols

Cardiac muscle tissue   Tissue of the heart
Carotene   Orange pigment
Cellular respiration   Combination of oxygen with food molecules to release the energy
Cementum   Material next to root that secures tooth in the jaw
Chromosome   A complete strand of DNA
Collagen   Flexible protein that builds the structure of bones
Compound fracture   One in which the bone punctures the skin
Connective tissue   Connects body parts
Crown   Visible part of the tooth
Cytosine   Base used in DNA, must be paired with guanine

DNA   Deoxyribonucleic acid, molecule containing genetic information
Dendrite   Part of the neuron that receives input
Dentin   Bonelike material surrounding pulp of the tooth
Deoxyribose   Sugar molecule forming the sides of the DNA molecule
Diastolic pressure   Blood pressure when the heart is at rest
Double helix   Shape of DNA; ladder that is twisted and compacted

Enamel   Hard protective covering on teeth
Endocrine system   Produces hormones
Epithelial tissue   Lines all body parts
Erector pili muscles   Tiny muscles in the skin attached to hairs

Eustachian tube   Opening between the middle ear and the throat
Excretory system   Removes wastes from the body
External respiration   Exchange of gases in the lungs

Forensic science   Study of items used in legal proceedings
Fracture   A break in a bone

Gastric juice   Enzymes and other chemicals secreted by the lining of the stomach
Gingiva   Gums
Guanine   Base used in DNA, must be paired with cytosine
Gustatory receptor cells   Cells that react with food molecules to produce electrical signals

Hormones   Chemical messengers to regulate body functions

Internal respiration   Exchange of gases between blood cells and tissue cells
Interneuron/Association neuron   Nerve cell that interprets input and generates output

Kidneys   Main organs of the excretory system

Ligament   Flexible cord-like material that connects bones together

Motor neuron   Nerve cell that carries a signal from the brain to a muscle
Muscle tissue   Contracts, for movement
Mutation   A mistake in the genetic code
Myelin   Material produced by the Schwann cells that provides insulation

Neck   Part of tooth entering the gums
Nerve tissue   Controls body activities

Occlusion   Correct teeth and jaw alignment or normal bite
Odorants   Light molecules that produce smell
Olfactory hairs   Dendrites of the nerves in the nasal cavity that chemically react with smell molecules
Orthodontics   The dental practice of straightening teeth and correcting jaw alignment

Pancreatic juice   Enzymes and other chemicals produced by the pancreas for digestion
Papillae   Taste buds

Periodontal ligament   Fiber between cementum and jawbone material
Pulp   Center of tooth containing nerves and blood vessels

Reproductive system   Produces children
Respiration   The exchange of oxygen and carbon dioxide
Retainer   Device to hold teeth in proper position
Root canals   Channels containing blood vessels in the tooth
Root   Part of tooth anchoring the tooth in the jaw

Salivary amylase   Enzyme that breaks down starch molecules
Schwann cells   Cells that cover and insulate the axon
Semicircular canals   Fluid-filled tubes that help determine balance
Sensory neuron   Nerve cell that receives input and carries it toward the brain
Simple fracture   One in which the bone does not puncture the skin
Smooth muscle tissue   Tissue designed for long, strong contractions
Sodium bicarbonate   Chemical produced by pancreas to neutralize stomach acid
Striated muscle tissue   Skeletal muscle tissue with striped appearance
Synovial fluid   Slippery fluid in the joint to facilitate smooth movement
Systolic pressure   Blood pressure when the ventricles of the heart contract

Thymine   Base used in DNA, must be paired with adenine

Universal donor   Blood type that can be donated to all other types
Universal recipient   Blood type that can receive all other types
Uterus   Womb

Vaccine   A substance that stimulates the immune system against a certain disease
Vitreous humor   Jelly-like substance in the middle of the eyeball

The World of Animals — Glossary

Abdomen   Back segment of an insect or other arthropod body
Airfoil   Shape that causes air to flow faster over a surface than under it creating lift
Amphibian   Animal that begins life breathing water and changes to be able to breathe air
Anal fin   Fin on the underside near the back of the fish
Antibiotic   Substance used to treat bacterial diseases
Apes   Primates without tails including chimps and gorillas
Arachnid   Animal with two body parts and eight legs
Arthropod   Animal with segmented legs or feet

Bacteria   Single-celled creatures without a defined nucleus
Baleen   Comb-like structures in a whale's mouth for straining food
Binocular vision   Eyes on the front of the head—each eye produces a slightly different view which when combined provides depth perception
Bivalve   Mollusk with two-part shell
Blowhole   Hole on the top of the head through which an aquatic mammal breathes

Cartilage   Flexible material replacing bone in some fish
Caudal fin   Tail fin
Cell membrane   Outer covering of cell, acts like skin
Centipede   Animal with segmented body with one pair of feet per segment
Cephalopod   Mollusk with a merged head and foot, and often no outer shell
Cephalothorax   Body part that is a combined head and thorax
Chitin   Starchy substance forming the exoskeleton
Chrysalis/Pupa   The stage in which the larva turns into an adult
Ciliate   Protist that moves using cilia
Cilia   Tiny hairs that cover a surface
Cloaca   Part of a bird's digestive system that releases waste
Cnidarians   Animals with hollow bodies and stinging tentacles
Cold-blooded   Animal that does not maintain a constant body temperature
Colubrid   Most common group of snakes

Complete metamorphosis   Change occurring in insects that look very different from their parents when they hatch
Compost   Decomposed material, fertilizer
Concertina movement   Moving by coiling and uncoiling
Constrictor   Snake that kills its prey by squeezing
Contour feathers   Feathers that cover a bird's body
Coral colony   A collection of thousands of coral connected together
Coral reef   A large collection of thousands of coral colonies connected together
Coral   Tiny cnidarians that grow a crusty shell around their bodies
Crop   Sac that releases food continuously into the bird's stomach
Crustacean   Animal that has two body parts and crusty exoskeleton
Cytoplasm   Liquid that fills a cell

Dorsal fins   Fins on the top of the fish
Down feathers   Fuzzy feathers providing insulation

Echinoderm   Sea creature with spiny skin, often has five legs
Endoskeleton   Internal skeleton
Esophagus   Tube between the mouth and stomach
Exoskeleton   Outer covering providing protection and support

Flagellate   Protist that moves using a flagellum
Flagellum   Whip-like structure that moves like a motor
Flatworm   Non-segmented worms with flat bodies
Flight feathers   Feathers that cover a bird's wings
Fluke   Tail fin on an aquatic mammal

Gastropod   Mollusk with one-part shell
Gills   Organs for removing oxygen from water
Gizzard   Rough organ to grind bird's food
Gullet   Opening that serves as a mouth

Head   Front segment of the insect body
Hibernation   A type of extended period of sleep

Incomplete metamorphosis   Change occurring in insects that look like their parents when they hatch

Insect   Animal with three body parts, six legs, wings, and antennae
Invertebrate   Animal without a backbone

Jacobson's organ   Special organ for smell found in snakes and some other reptiles
Joey   Immature marsupial

Keratin   Material that forms hair, fingernails, and baleen

Larva/Larval stage   Early stage of an animal that undergoes metamorphosis
Lateral undulation   Moving in sideways waves

Mammal   Warm-blooded animal with fur and mammary glands
Mammary glands   Glands that secrete milk for feeding young
Mantle   Organ that secretes a substance that forms a shell
Marsupial   Mammal with a pouch for carrying developing young
Medusa   Adult stage of a jellyfish's lifecycle when it has a bell-shaped body
Metamorphosis   A significant change in form
Millipede   Animal with segmented body with two pairs of feet per segment
Mitochondria   Cell's power plants
Mollusk   Soft-bodied invertebrate with a muscular foot and usually a shell
Moneran   Microorganisms without a nucleus, including different bacteria
Myriapod   Animal with many feet, specifically centipedes and millipedes

New World monkey   Monkeys that live in the Western Hemisphere, have a prehensile tail
Nictitating membrane   Clear eyelids that protect a reptile's eyes
Nocturnal   Active at night
Nucleus   Control center of the cell
Nymph   Immature insect that experiences incomplete metamorphosis

Old World monkey   Monkeys that live in the Eastern Hemisphere, do not have a prehensile tail

Parasite   Animals that take nutrients from a living host
Pectoral fins   Front fins used for angling up and down

Pelvic fins   Fins on bottom of fish in center of body
Planula   Worm-like stage in a jellyfish's lifecycle before it becomes a polyp
Polyp   The stage in a cnidarian's life when it has hollow body with tentacles
Pore   Small openings or holes
Preening   Running the feather through the beak to re-hook the barbs
Prehensile tail   One which has the ability to grasp
Primate   Mammal with five fingers, five toes, and binocular vision
Protist   A diverse group of simple creatures with a nucleus
Pseudopod   Foot or finger-like projection of a cell

Rectilinear movement   Moving by contracting and stretching to move in a straight line
Regenerate   To regrow a lost body part
Rostrum   Beak of a dolphin
Roundworm   Non-segmented worms with round bodies

Sarcodine   Protist that moves using pseudopods
Scavengers   Animals that eat dead plants or animals
Segmented body   Animal with distinct sections of its body
Segmented worm   Worm with rings or segments to its body
Side winding   Moving forward at an angle by moving sideways at the same time
Spinnerets   Organs which produce silky thread
Sponge   Simple animal with many pores
Swim bladder   Balloon-like sac used for buoyancy
Symbiotic relationship   Two or more creatures living in a mutually beneficial way

Tadpole/Pollywog   The larva or infant form of an amphibian
Talons   Claw-like feet
Thorax   Middle segment of the insect body

Vaccine   Substance that causes a body to build immunity to disease
Vacuole   Storage area in a cell
Venomous   Snakes that have a poisonous bite
Vertebrae   Small bones that protect the spinal cord
Vertebrate   Animal with a backbone
Virus   Sub-microscopic agent that causes disease

Warm-blooded   Animal that maintains a constant body temperature

The World of Animals — Challenge Glossary

Abomasum   Fourth chamber of a ruminant's stomach
Antibiotic resistant bacteria   Bacteria that are not killed by certain antibiotics
Asymmetrical   Having no symmetry

Bilateral symmetry   Can be divided symmetrically by only one lateral line
Bioluminescence   Process producing light in an animal through chemical reactions
Biomimetics   Study of living creatures for human technology
Buoyant   Able to float

Carapace   Top part of a turtle shell
Casque   Large bony structure on the head of a cassowary bird
Ceratopsians   Horned dinosaurs
Chemosynthesis   Process of converting chemicals into food
Circular canal   Central canal pumping water to the ray canals
Compound eye   Eye with multiple lenses
Counter-current exchange   Air and blood flow in opposite directions through the lungs
Cud   The food that is regurgitated for more chewing

Digitigrade   Walking on the base or flats of the toes

Echolocation   Sonar used by animals for communication

Fiber optics   Use of tiny glass tubes to transmit light

Hemoglobin   Substance that turns bright red in the presence of oxygen
Hydrothermal vent   Area on ocean floor where super-heated water flows out

Lateral line   Series of nerves covering the head and sides of a fish
Leptoid scales   Scales on bony fish that grow as the fish grows

Madreporite   Openings through which water enters the water vascular system

Olfactory lobe   Part of the brain responsible for the sense of smell

Omasum   Third chamber of a ruminant's stomach
Open circulatory system   One with no blood vessels to carry the blood
Optic lobe   Part of the brain responsible for the sense of sight

Placoid scales   Scales on cartilaginous fish that do not grow
Plantigrade   Walking on the soles of the feet
Plasmodium   Sporozoan that causes malaria
Plastron   Bottom part of a turtle shell
Plated dinosaurs   Dinosaurs with large plates along their backs
Plume   End of a tubeworm's body

Radial symmetry   Can be divided symmetrically by any lateral line through the center of the circle
Ray canals   Tubes carrying water to the rays of the starfish
Reticulum   Second chamber of a ruminant's stomach
Rorqual   Whales with grooved expandable throats
Rumen   First chamber of a ruminant's four chambered stomach
Ruminant   Animal that regurgitates and rechews its food

Sauropods   Large dinosaurs with long necks and tails
Siphonophore   A colony of cnidarians living together to form one organism
Spherical symmetry   Can be divided symmetrically by any line through the center of the body
Spiracles   Openings in an insect's side for air flow
Sporozoan   Protist that produces spores
Stance   The way an animal walks on its feet

Theropods   Meat-eating dinosaurs
Tube feet   Rows of tubes on the underside of each starfish ray
Tube worm   Worm that thrives near hydrothermal vents

Ungulates   Animals with an unguligrade stance
Unguligrade   Walking on the tips of the toes, usually with hooves
Urticating hairs   Barbed hairs on a spider that produce irritation in enemies

Water vascular system   Series of tubes that carry water throughout the starfish's body

Daily Lesson Plans

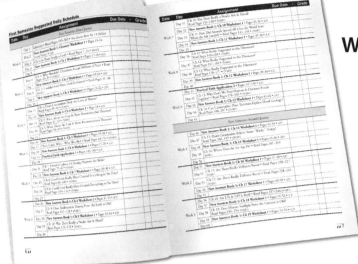

WE'VE DONE THE WORK FOR YOU!

PERFORATED & 3-HOLE PUNCHED
FLEXIBLE 180-DAY SCHEDULE
DAILY LIST OF ACTIVITIES
RECORD KEEPING

"THE TEACHER GUIDE MAKES TH
SO MUCH EASIER AND TAKES
GUESS WORK OUT OF IT FO

HOMESCHOO

Master Books® Homeschool Curricu

Faith-Building Books & Resources
Parent-Friendly Lesson Plans
Biblically-Based Worldview
Affordably Priced

Master Books® is the leading publisher of books and resourc based upon a Biblical worldview that points to God as our C
Now the books you love, from the authors you trust like Ken Ham, Michael Farris, Tommy Mitchell, and many more are available as a homeschool curriculum.